X-Machines for Agent-Based Modeling

FLAME Perspectives

CHAPMAN & HALL/CRC
COMPUTER and INFORMATION SCIENCE SERIES

Series Editor: Sartaj Sahni

PUBLISHED TITLES

ADVERSARIAL REASONING: COMPUTATIONAL APPROACHES TO READING THE OPPONENT'S MIND
Alexander Kott and William M. McEneaney

COMPUTER-AIDED GRAPHING AND SIMULATION TOOLS FOR AUTOCAD USERS
P. A. Simionescu

DELAUNAY MESH GENERATION
Siu-Wing Cheng, Tamal Krishna Dey, and Jonathan Richard Shewchuk

DISTRIBUTED SENSOR NETWORKS, SECOND EDITION
S. Sitharama Iyengar and Richard R. Brooks

DISTRIBUTED SYSTEMS: AN ALGORITHMIC APPROACH, SECOND EDITION
Sukumar Ghosh

ENERGY-AWARE MEMORY MANAGEMENT FOR EMBEDDED MULTIMEDIA SYSTEMS:
A COMPUTER-AIDED DESIGN APPROACH
Florin Balasa and Dhiraj K. Pradhan

ENERGY EFFICIENT HARDWARE-SOFTWARE CO-SYNTHESIS USING RECONFIGURABLE HARDWARE
Jingzhao Ou and Viktor K. Prasanna

FROM ACTION SYSTEMS TO DISTRIBUTED SYSTEMS: THE REFINEMENT APPROACH
Luigia Petre and Emil Sekerinski

FUNDAMENTALS OF NATURAL COMPUTING: BASIC CONCEPTS, ALGORITHMS, AND APPLICATIONS
Leandro Nunes de Castro

HANDBOOK OF ALGORITHMS FOR WIRELESS NETWORKING AND MOBILE COMPUTING
Azzedine Boukerche

HANDBOOK OF APPROXIMATION ALGORITHMS AND METAHEURISTICS
Teofilo F. Gonzalez

HANDBOOK OF BIOINSPIRED ALGORITHMS AND APPLICATIONS
Stephan Olariu and Albert Y. Zomaya

HANDBOOK OF COMPUTATIONAL MOLECULAR BIOLOGY
Srinivas Aluru

HANDBOOK OF DATA STRUCTURES AND APPLICATIONS
Dinesh P. Mehta and Sartaj Sahni

PUBLISHED TITLES CONTINUED

HANDBOOK OF DYNAMIC SYSTEM MODELING
Paul A. Fishwick

HANDBOOK OF ENERGY-AWARE AND GREEN COMPUTING
Ishfaq Ahmad and Sanjay Ranka

HANDBOOK OF GRAPH THEORY, COMBINATORIAL OPTIMIZATION, AND ALGORITHMS
Krishnaiyan "KT" Thulasiraman, Subramanian Arumugam, Andreas Brandstädt, and Takao Nishizeki

HANDBOOK OF PARALLEL COMPUTING: MODELS, ALGORITHMS AND APPLICATIONS
Sanguthevar Rajasekaran and John Reif

HANDBOOK OF REAL-TIME AND EMBEDDED SYSTEMS
Insup Lee, Joseph Y-T. Leung, and Sang H. Son

HANDBOOK OF SCHEDULING: ALGORITHMS, MODELS, AND PERFORMANCE ANALYSIS
Joseph Y.-T. Leung

HIGH PERFORMANCE COMPUTING IN REMOTE SENSING
Antonio J. Plaza and Chein-I Chang

HUMAN ACTIVITY RECOGNITION: USING WEARABLE SENSORS AND SMARTPHONES
Miguel A. Labrador and Oscar D. Lara Yejas

IMPROVING THE PERFORMANCE OF WIRELESS LANs: A PRACTICAL GUIDE
Nurul Sarkar

INTEGRATION OF SERVICES INTO WORKFLOW APPLICATIONS
Paweł Czarnul

INTRODUCTION TO NETWORK SECURITY
Douglas Jacobson

LOCATION-BASED INFORMATION SYSTEMS: DEVELOPING REAL-TIME TRACKING APPLICATIONS
Miguel A. Labrador, Alfredo J. Pérez, and Pedro M. Wightman

METHODS IN ALGORITHMIC ANALYSIS
Vladimir A. Dobrushkin

MULTICORE COMPUTING: ALGORITHMS, ARCHITECTURES, AND APPLICATIONS
Sanguthevar Rajasekaran, Lance Fiondella, Mohamed Ahmed, and Reda A. Ammar

PERFORMANCE ANALYSIS OF QUEUING AND COMPUTER NETWORKS
G. R. Dattatreya

THE PRACTICAL HANDBOOK OF INTERNET COMPUTING
Munindar P. Singh

SCALABLE AND SECURE INTERNET SERVICES AND ARCHITECTURE
Cheng-Zhong Xu

PUBLISHED TITLES CONTINUED

SOFTWARE APPLICATION DEVELOPMENT: A VISUAL C++®, MFC, AND STL TUTORIAL
Bud Fox, Zhang Wenzu, and Tan May Ling

SPECULATIVE EXECUTION IN HIGH PERFORMANCE COMPUTER ARCHITECTURES
David Kaeli and Pen-Chung Yew

TRUSTWORTHY CYBER-PHYSICAL SYSTEMS ENGINEERING
Alexander Romanovsky and Fuyuki Ishikawa

VEHICULAR NETWORKS: FROM THEORY TO PRACTICE
Stephan Olariu and Michele C. Weigle

X-Machines for Agent-Based Modeling: FLAME Perspectives
Mariam Kiran

X-Machines for Agent-Based Modeling

FLAME Perspectives

Mariam Kiran

CRC Press
Taylor & Francis Group
Boca Raton   London   New York

CRC Press is an imprint of the
Taylor & Francis Group, an **informa** business

A CHAPMAN & HALL BOOK

CRC Press
Taylor & Francis Group
6000 Broken Sound Parkway NW, Suite 300
Boca Raton, FL 33487-2742

Printed and bound in Great Britain by
TJ International Ltd, Padstow, Cornwall

To my family,
FLAME contributors and
its users.

Contents

Foreword   xiii

Preface   xvii

List of Figures   xix

List of Tables   xxv

FLAME Contributors   xxvii

1 Setting the Stage: Complex Systems, Emergence and
   Evolution   1

   1.1 Complex and Adaptive Systems   3
   1.2 What Is Chaos?   4
   1.3 Constructing Artificial Systems   5
   1.4 Importance of Emergence   6
   1.5 Dynamic Systems   6
   1.6 Is There Evolution at Work?   7
      1.6.1 Adaptation   8
   1.7 Distributing Intelligence?   9
   1.8 Modeling and Simulation 10
      1.8.1 Research Examples 12

2 Artificial Agents   17

   2.1 Intelligent Agents 18
      2.1.1 "Can Machines Think?" 19
   2.2 Engineering Self-Organizing Systems 21
      2.2.1 Bring in the Agents 22
      2.2.2 Characteristics of Agent-Based Models 23
   2.3 Agent-Based Modeling Frameworks 33
   2.4 Adaptive Agent Design 37
   2.5 Mathematical Foundations 38
   2.6 Objects or Agents? 39
   2.7 Influence of Other Research Areas on ABM 40

3   Designing X-Agents Using FLAME   **43**

   3.1   FLAME and Its X-Machine Methodology   44
      3.1.1   Transition Functions   47
      3.1.2   Memory and States   47
   3.2   Using Agile Methods to Design Agents   48
      3.2.1   Extension to Extreme Programming   51
   3.3   Overview: FLAME Version 1.0   51
   3.4   Libmboard (FLAME message board library)   54
      3.4.1   Compiling and Installing Libmboard   55
      3.4.2   FLAME's Synchronization Points   57
   3.5   FLAME's Missing Functionality   58

4   Getting Started with FLAME   **61**

   4.1   Setting Up FLAME   62
      4.1.1   MinGW   63
      4.1.2   GDB GNU Debugger   63
      4.1.3   Dotty as an Extra Installation   64
   4.2   Messaging Library: Libmboard   64
   4.3   How to Run a Model?   65
   4.4   Implementation Details   65
   4.5   Using Grids   68
   4.6   Integrating with More Libraries   69
   4.7   Writing a Model - Fox and Rabbit Predator Model   71
      4.7.1   Adding Complexity to Models   72
      4.7.2   XML Model Description File   72
      4.7.3   C Function   76
      4.7.4   Additional Files   81
      4.7.5   0.xml File   83
   4.8   Enhancing the Environment   84
      4.8.1   Constant Variables   84
      4.8.2   Time Rules   84

5   Agents in Social Science   **87**

   5.1   Sugarscape Model   92
      5.1.1   Evolution from Bottom-Up   93
      5.1.2   Distribution of Wealth   94
      5.1.3   Location Is Important!   95
      5.1.4   Find Agents around Me   104
      5.1.5   Handle Multiple 'Eaten' Requests   105
      5.1.6   Change Starting Conditions   105
   5.2   Modeling Social Networks   107
      5.2.1   Set Up a Recurring Function   112

     5.2.2   Assigning Conditions with Functions 113

     5.2.3   Using Dynamic Arrays and Data Structures 113

     5.2.4   Creating Local Dynamic Arrays 114

  5.3 Modeling Pedestrians in Crowds 114

     5.3.1   Calculate Movement toward Other Agents 116

     5.3.2   Entering and Exiting Agents 118

6 Agents in Economic Markets and Games   **121**

  6.1 Perfect Rationality versus Bounded Rationality 125

  6.2 Modeling Multiple Shopper Behaviors 126

  6.3 Learning Firms in a Cournot Model 129

     6.3.1   Genetic Programming with Agents 143

     6.3.2   Filtering Messages in Advance 150

     6.3.3   Comparing Two Data Structures 151

  6.4 A Virtual Mall Model: Labor and Goods Market Combined . 152

  6.5 Programming Games 159

     6.5.1   Nash Equilibrium 160

     6.5.2   Evolutionary Game Theory 161

     6.5.3   Evolutionary Stable State 162

     6.5.4   Game Theory versus Evolutionary Game Theory . . . 162

     6.5.5   Continuous Strategies 163

     6.5.6   Red Queen and Equilibrium 163

  6.6 Learning in an Iterated Prisoner's Dilemma Game 164

  6.7 Multi-Agent Systems and Games 173

7 Agents in Biology   **175**

  7.1 Example Models 176

     7.1.1   Molecular Systems Models 176

     7.1.2   Tissue and Organ Models 179

     7.1.3   Ecological Models 182

     7.1.4   Industrial Applications of Agent-Based Modeling with
             FLAME 183

  7.2 Modeling Epithelial Tissue 184

     7.2.1   Merging with Other Toolkits 185

  7.3 Modeling Drosophila Embryo Development 187

     7.3.1   Stochastic Modeling 188

     7.3.2   Converting to an Agent-Based Model 188

     7.3.3   Find Optimum Model Settings 196

  7.4 Output Files for Analysis 198

  7.5 Modeling Pharaoh's Ants (*Monomorium pharaonis*) 202

  7.6 Model Drug Delivery for Cancer Treatment 224

     7.6.1   Using Multiple Outputs 234

8   Testing Agent Behavior   **237**

   8.1   Unit and System Testing 237
   8.2   Statistical Testing of Data 239
   8.3   Statistics Testing on Code 243
   8.4   Testing Simulation Durations 244

9   FLAME's Future   **247**

   9.1   FLAME to FLAME GPU 247
      9.1.1   Visualizing Is Easy in FLAME GPU 273
      9.1.2   Utilizing Vector Calculations 276
   9.2   Commercial Applications of FLAME 276

Bibliography   **283**

Index   **299**

Foreword

Simulation is a powerful tool, that allows domain experts to test their theories as safe virtual experiments. But as the systems being modeled grow and become complex, with many interacting elements, the code also becomes extremely complex. Whether it be modeling an ant colony, or human interactions in economic systems, these problems not only help the domain experts, but also require immense effort from computer scientists. A multitude of computer science techniques are involved such as how to design models, build code, simulate and analyze data. Agent-based modeling is an example of simulation technique, which can help researchers deviate from stochastic and differential equations, to more granular approaches of building models based on interactions.

Agent-based models have shown applications in various fields such as biology, economics and social sciences. Over the years, multiple agent-based modeling frameworks have been produced, allowing experts with non-computing background to easily write and simulate their models. However, most of these models are limited by the capability of the framework, time it takes for a simulation to finish, or handling the massive amounts of data produced. FLAME (Flexible Large-scale Agent-based Modeling Environment) was produced at the University of Sheffield, and developed through the years, with multiple grants and projects from biology, sociology and economics. As a challenge, it was able to produce an economic agent-based model, EURACE, consisting of three markets integrated together, which had never been done before.

This book contains a comprehensive summary of the field and how concepts of X-machines can be stretched across multiple fields to produce agent models. It has been written with several audiences in mind. First, it is organized as a collection of models, with detail descriptions of how models can be designed, especially for beginners in agent-based models. A number of theoretical aspects of software engineering and how they relate to agent-based models have been discussed for students interested in software engineering and parallel computing. Finally, it is intended as a guide to developers from biology, economics and sociologists, who want to explore how to write agent-based models for their research area. By working through model examples provided, anyone should be able to design and build their agent-based models and deploy them on their machines. With FLAME, they can easily increase the agent number and run models on parallel computers, in order to save on simulation complexity and waiting time for results.

Because the field is so large and active, this book does not aim to cover all aspects of agent-based modeling and its research challenges. The models are presented to aid researchers with capability, on how they can build complex agent functions for their models. This book will give a good feeling, making researchers confident on writing their agent-based models and the complexities which go behind it. Finally, the book should convince anyone of the advantage of using agent-based models in their simulation experiments, providing the case to move away from differential equations and build more reliable, close to real models.

It is important to acknowledge all the people who have contributed to the book and the FLAME framework, through their models, images and code, maturing FLAME into an independent toolkit. It is a product of many years of research, learning, ideas and collective efforts. Many people have come together to make this book a possibility. The author acknowledges that FLAME is a product on various developers and researchers, part of the FLAME family over the years. Developed as part of Simon Coakley's PhD thesis, the framework has matured into a commercial tool, with very real world applications. Lastly, I would like to thank Professor Mike Holcombe for his leadership, imagination and limitless ideas during the years for FLAME's growth and also encouragements for putting this book together, for summarizing FLAME efforts.

About the Author

Dr. Mariam Kiran is a well-recognized researcher in agent-based modeling, high performance simulations and cloud computing. She has published numerous papers in these fields, both, in theory and practical implementations, exploiting grid and cloud ecosystems for improving computational performance for multi-domain research. She has an extensive record of research collaborations across the world, serving as a board member for Complex Systems research in CoMSES, and several joint projects funded by European Research and UK Engineering Council. She is also active in education research of software engineering in team building and writing software for simulations.

Mariam Kiran received her PhD in Computer Science from University of Sheffield, Sheffield UK in 2010. She is currently involved in many projects at Lawrence Berkeley National Labs, California, optimizing high performance computing problems across various disciplines. Prior to this, she was working as an Associate Professor at University of Bradford, leading the Cloud Computing research in the School.

The author's research focuses on learning and decentralized optimization of system architectures and algorithms for high performance computing, using underlying networks and Cloud infrastructures. She has been exploring various platforms such as HPC grids, GPUs, Cloud and SDN-related technologies. Her work optimizes quality of service of applications, parallelization performance and solves complex data intensive problems such as large-scale complex simulations.

For the Reader

This book is intended primarily as a textbook for researchers and developers exploring uses of agent-based modeling and of Flame. Certain aspects of the book are specifically designed to help researchers:

- Code examples of many agent-based models from different disciplines. These make arguments that any kind of real-world model can be converted into a simulation model, using the same principles for building and agent-based model.

- Mathematical use of simulations. The use of maths formulas and data extraction shows how simulations also follow the same rules of real world physics and geometry, when real-world problems are being adapted in simulations. If these are modeled correctly, the model will be an accurate representation of the problem.

- Using models to test theories in simulation environment. The book gives examples that any complex system can be modeled as a simulation. Agent-based models are the best manner to model these, instead of traditional differential equations, as they allow more complex individual behavior to be modelled from bottom up rather than top-to-bottom.

The book assumes that the readers have some knowledge of programming languages such as C, Java, Algorithm design and some knowledge of state machine models. This is useful to link theory to simulation constructions. The book explains in detail how X-machines are being adapted for agent design.

Preface

The world seems to be a more and more complex place and trying to understand this complexity is a serious challenge for the future. Whether it is the fundamental basis for life or the increasingly global nature of society, the need to be able to model, predict and explore these phenomena is becoming increasingly important.

Alongside the massive increases in the data that technology and society are generating fundamental questions of,

- What do all these data mean?

- How can we understand all the interconnections that underlie the data?

- Can we model these systems and predict what they may do in the future?

- And build on this knowledge in order to understand and control our world better?

- And create sensible policies for deciding the future?

For many types of systems, be they molecular process inside a cell or the manifestations of economic activity, it is being realized that the old ways of modeling and predicting their behavior are no longer useful. We can no longer assume that a cell is a bag of randomly moving chemicals (molecules) since the intimate interactions between individual chemicals and where in the cell these interactions take place are of fundamental importance. Similarly, the old assumptions that economics is based on generalized rational behavior and that markets are inherently stable have been discredited by the recent economic crises that have beset the world.

A new approach is needed and this is now feasible because technology now allows for highly detailed modeling of these complex systems. This book exemplifies one of the most successful approaches to modeling and simulating this new generation of complex systems.

FLAME was designed to make the building of large-scale complex systems models straightforward and the simulation code that it generates is highly efficient and can be run on any modern technology. FLAME was the first such platform that ran efficiently on high performance parallel computers (or HPC) and a version for NVidia GPU-technology (Graphical Processing Unit) is also available.

Writing complex simulation code is an error-prone process and rarely meets the standards required for best practice software engineering. This is true of many Agent-based Modeling (ABM) platforms. For people to believe the results of a simulation model and the model-building process it needs to be transparent. Journals and others are demanding much more information about the details of the models. FLAME addresses these issues by providing a basic notation for describing agents and a robust translation process that automatically generates executable code. FLAME was built using the latest software engineering approaches.

At its heart, and the reason why it is so efficient and robust, is the use of a powerful computational model 'Communicating X-machines' which is general enough to cope with most types of modeling problem. As well as being increasingly important in academic research FLAME is now being applied in industry in many different application areas.

This book describes the basics of FLAME and is illustrated with numerous examples.

Professor Mike Holcombe

List of Figures

1.1   Emergence in complex systems. cf. [116].   2

1.2   Examples of complex adaptive systems, their models and common characteristics. cf. [171].   4

1.3   Bifurcation diagram in a logistic map. Adapted from [122]. .   5

1.4   Examples of Karl Sims's creatures. cf. [180].   8

1.5   Modeling process in biology simulations. cf. [107].   12

1.6   Various time steps showing ant colonies finding and forming routes to food sources. cf. [23].   13

1.7   A system working in the environment. The system is composed of three elements working together to make the system work efficiently. Output produces a feedback, that produces change in the system as time progresses.   14

1.8   Research areas of 'Scientific Computing' and 'Distributed Computing' have a close relationship in agent-based modeling.   15

2.1   Scientific method. cf. [61].   18

2.2   Separate researches in AI.   19

2.3   State and X-machine diagrams.   20

2.4   Your mind designed for CogAff Project. cf. [183].   24

2.5   Weak and strong notions of agent actions. Cf. [205].   25

2.6   Program represented as a tree and a string. cf. [50].   28

2.7   An agent can represent a single strategy or multi-strategies.   31

2.8   Evolvability of programs.   32

2.9   Nested hierarchy of swarms.   34

3.1   Block diagram of FLAME. cf. [76].   44

3.2   State and X-machine diagrams of an ant foraging for food. cf. [104].   46

3.3   FLAME uses strict X-machine architecture - Memory, Functions, States and Messages.   48

3.4   Incorporating agile methodology in agent models. Modified from [20].   49

3.5   Agile agent development process.   50

3.6   Structure of basic agent. Agents represent any individual such as a household, an ant or a firm.   52

3.7   Two X-machine agents communicating through a message board. The message board library (Libmboard) saves current active messages during the simulation time step.   52

3.8   One iteration with two agents, each with two functions. . . .   53

3.9   Transition functions perform on memory variables.   54

3.10   Serial versus parallel execution of agents.   55

3.11   Distributed memory and synchronization.   55

3.12   Using filters and iterators to quicken message parsing for agents.   56

3.13   Simulation times across multiple processor nodes.   56

3.14   Timeline showing when the synchronization point occurs when messages interact with functions.   58

4.1   Block diagram of the Xparser, the FLAME simulation component. Blocks in blue are files automatically generated. The green blocks are modeler's files.   62

4.2   FLAME software blocks.   63

4.3   Flow diagram for the simulation describing agents, its functions and communications.   71

4.4   Flow diagram for simulation describing agents, their functions and communications between the agents with complexity. .   73

4.5   Iteration files with updated agent memory results.   83

5.1   Snapshot of game of life during a simulation. Adapted from [141].   89

5.2   Initial distribution of sugar (left) and with agents (right). Adapted from [21].   90

5.3   Agent perception. They can see north, south, east and west.   90

5.4   During the simulation, agents move to high sugar concentration areas. Adapted from [21].   91

5.5   Relationships emerged between rich and poor agents. The middle agents behaved like banks.   94

5.6   Wealth distribution among agents, with initial random sugar distribution. cf. [21].   95

5.7   View of a citizen agent in FLAME Sugarscape.   96

5.8   Timeline of the basic FLAME Sugarscape model.   97

5.9   Three different initial settings for simple Sugarscape experiment. The citizen agents are represented by *red* dots and *green* dots represent sugar agents in the scenario.   106

5.10   Sugar collected for random initial agent distribution.   107

5.11   Distribution of captured sugar.   108

5.12   Evolution of networks in a simulation. Adapted from [151]. .   109

5.13   Evolved centralization and density in networks. Adapted from [151].   110

5.14   Using vector equations to calculate resulting movement. . .   117

5.15 Agents walking in the scene. 119

6.1 A black box represents an economic model where only inputs and outputs are known and little is known about what goes on inside. 122
6.2 Groups in economic systems. 122
6.3 Replacing the black box with agents. 123
6.4 Different shoppers in the same simulation. 128
6.5 Five firms producing a particular output of the same product in the scenario. 129
6.6 A supply-demand curve. 131
6.7 Time line of the various activities in a simple Cournot model. 132
6.8 Firm reaction curves in a duopoly model. A duopoly market is a market with only two acting firms. Adapted from [5]. . 133
6.9 An evolutionary model. Each firm has its own strategy base which after every simulation is updated using genetic algo-rithms. Adapted from [5]. 134
6.10 All firms have a strategy base in their memory. 136
6.11 How a strategy base looks in firm's memory. 136
6.12 How crossover works in Cournot model. Strategy A, B, C and D represent numerical values of bit strings. 137
6.13 How mutation works in Cournot model. 137
6.14 The system contains four agents (three firms and one system demand, responsible for assessing the product price). 137
6.15 Quantity and profits of three firms. 139
6.16 Price and strategy space in evolution. 139
6.17 Average price with crossover rate 0.1, 0.5 and multiple muta-tion rates. 141
6.18 Average price with crossover rate 0.7 and multiple mutation rates. 142
6.19 Mall capitals and worker savings. 153
6.20 People wages and unemployment. 155
6.21 Mall strategies and costs. 155
6.22 Stategraph for virtual mall experiment. 156
6.23 Example automaton for prisoner's dilemma strategies. . . . 167
6.24 Finite state machine of eight states representing a prisoner's dilemma strategy. cf. [81]. 168
6.25 Example of automaton represented by Table 6.8. 169
6.26 Strategy database of ten strategies in player memory. . . . 170
6.27 One state in a strategy. 170
6.28 Two strategies acting as parents. 171
6.29 Two children resulting from crossover of parents, at crossover point state number 1 and state length 4. 171
6.30 Mutant child of Parent-1 at mutation point state number 1 and state length 4. 172

6.31   Score is payoff returned playing IPD game.   172

7.1   Industrial applications of FLAME.   184
7.2   Comparing real and simulated data of wound healing in 2D.
       Adapted from [192].   185
7.3   3D model snapshots of wound healing at different time steps
       of the simulation.   186
7.4   Calling Copasi from FLAME C code.   187
7.5   Movement of proteins within a Drosophila embryo. A struc-
       tured view.   189
7.6   Agent activities during one iteration.   190
7.7   Bicoid concentration profiles jointly in A-P axis and develop-
       mental time, shows a deterministic model output as an average
       value of stochastic model (A), to one stochastic simulation (B)
       and the results of one agent-based model run (C).   198
7.8   One realization of stochastic simulation using Gillespie Al-
       gorithm at different time points: 60 (A), 100 (B), 144 (C)
       and 180 (D) min. Blue histograms show number of Bicoid
       molecules along anterior and posterior axis in embryo. Red
       lines show average amount of molecules from deterministic
       reaction diffusion model. Bicoid intensity at 144 min (C) is
       the peak stage and will degrade after mRNA regulation. Red
       histograms show number of Bicoid molecules along anterior
       and posterior axis in embryo resulting from average 20 runs
       of the agent-based model simulation.   199
7.9   The agent-based modeling simulation result with stochastic
       model. The circle shows missing data points in agent-based
       results using same initial settings in both models.   200
7.10   Zoom in to find shortest possible error between simulated re-
       sults in agent-based, stochastic and original datasets.   200
7.11   Using shortest possible error between the simulated results.   201
7.12   Ant simulation.   204
7.13   Sequential trails for drug therapy.   225
7.14   Testing drug effect on cancer cells.   235
7.15   Multiple views of the same cancer model.   236

8.1   Testing low and high level functions.   238
8.2   Screenshot of Weka analyzing cancer output.   239
8.3   Plots of no-hope cells with correct and incorrect codes. . . .   243
8.4   Simulation times of models.   245

9.1   Diagram of ABM-based decision support system.   277
9.2   Patient flow in green zone versus resource usage.   278
9.3   Part of a Concoursia simulation of a London main station. .   280

9.4 A heat map of the station showing overcrowding (red) and a graphical output over time. 281

9.5 Modeling the impact of a car display and a pop-up kiosk in a station concourse. 281

List of Tables

1.1   Examples of research carried out in complex adaptive systems. Adapted from Schuster [171].   3

2.1   Comparison of agent-based modeling frameworks.   33

3.1   Simulation times across multiple processors in HPC grids [76].   56

4.1   Model parameters for fox and rabbit example.   72

5.1   Global variables used in FLAME Sugarscape.   96
5.2   FLAME Sugarscape model.   97
5.3   Results of skewness and kurtosis measures in three experiments.   108
5.4   Social network model specifications.   110

6.1   Five big ideas that distinguish complexity economics [21]. .   124
6.2   Evolving Cournot characteristics for each firm.   135
6.3   Numerical values in Cournot experiment.   138
6.4   Difference between Nash equilibrium and evolutionary stable state.   164
6.5   Prisoner sentences in PD game.   165
6.6   Payoff matrix in PD game, where R=3, S=0, T=5, P=1. . .   165
6.7   All defect strategy (Player 1) playing against a tit-for-tat strategy (Player 2).   167
6.8   Example of a three state machine represented by automaton.   169
6.9   Numerical values in FLAME-IPD experiment.   172

7.1   Comparing building simulations in MATLAB and FLAME.   197
7.2   Different initial value setting for the Bicoid ABM.   198

FLAME Contributors

FLAME is a toolkit for agent-based modeling, developed through multiple projects and teams. These people come from backgrounds in Computer Science, such as software engineering, high performance computing, graphics and also biology, economy and social science.

Mike Holcombe
University of Sheffield
Sheffield, UK

Simon Coakley
University of Sheffield
Sheffield, UK

Chris Greenough
STFC R.A. Laboratory
Didcot, UK

Rod Smallwood
University of Sheffield
Sheffield, UK

David Worth
STFC R.A. Laboratory
Didcot, UK

Shawn Chin
STFC R.A. Laboratory
Didcot, UK

Phil Mcminn
University of Sheffield
Sheffield, UK

Susheel Varma
University of Sheffield
Sheffield, UK

Salem Adra
University of Sheffield
Sheffield, UK

Mark Pogson
University of Sheffield
Sheffield, UK

Mesude Bicak
University of Sheffield
Sheffield, UK

Afsaneh M.Dizaji
University of Sheffield
Sheffield, UK

Daniela Romano
University of Sheffield
Sheffield, UK

Paul Richmond

University of Sheffield
Sheffield, UK

Twin Karmakharm
University of Sheffield
Sheffield, UK

Hao Bai
University of Sheffield
Sheffield, UK

Sander van der Hoog
University of Bielefeld
Bielefeld, Germany

Mark Burkitt
University of Sheffield
Sheffield, UK

Dawn Walker
University of Sheffield
Sheffield, UK

Mariam Kiran
University of Sheffield
Sheffield, UK

Chapter 1

Setting the Stage: Complex Systems, Emergence and Evolution

1.1	Complex and Adaptive Systems	3
1.2	What Is Chaos?	4
1.3	Constructing Artificial Systems	5
1.4	Importance of Emergence	6
1.5	Dynamic Systems	6
1.6	Is There Evolution at Work?	7
	1.6.1 Adaptation	8
1.7	Distributing Intelligence?	9
1.8	Modeling and Simulation	10
	1.8.1 Research Examples	12
	Natural Systems	12
	Control Engineering	13
	Cellular Automata	14
	Agent-Based Models	14

COMPLEX SYSTEMS are composed of many interconnected elements, working individually, but producing an overall global system behavior. The fundamental desire to study how these complex systems behave comes from various multifaceted disciplines such as biology, economics or even social sciences. Examples include large ant colonies (composed of individual ants cooperating to exploit available food sources), the human nervous system (composed of tiny neurons sending and receiving signals in the human body) or social structures (such as communication networks). Depending on the system being studied, individuals behave in organized (or disorganized) ways, leading to unpredictable overall system behavior. This phenomenon, referred to as *emergent behavior*, is a direct consequence of individual behaviors inside the system and their interactions among each other.

Engineering projects have taken inspiration from complex natural systems to build better and reliable infrastructures. Understanding how cities survive and how crowds behave are key elements in designing buildings or studying how economies work.

Agent-based modeling (ABM) is a unique modeling technique that allows a one-to-one mapping to natural systems. Modelers understand complex systems, how they are composed of multiple individuals and their interactions

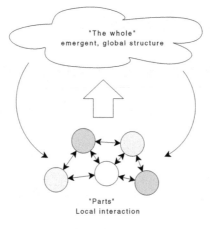

FIGURE 1.1: Emergence in complex systems. cf. [116].

among themselves and the environment. Writing agent-based models draws inspiration from parallel computation, software engineering, data analysis and simulation, to achieve reliable simulation models as virtual complex systems. This book aims to study and provide readers with principals involved in building and writing agent-based models from a software engineering perspective. Presenting itself primarily as a modeling and simulations tool, the book covers computational challenges of software engineering, parallelization, verification and validation, all of which are issues for computer and other scientists when developing reliable agent-based models. To explain details from a software engineering perspective, we focus on an established agent-based modeling framework, FLAME, as a guide to understand and build ABM approaches. By discussing the range of projects and computational complexities it has faced in the research area, various computational challenges are discussed from model conception, building, execution and testing, in fields of biology, social networks and economics.

Complex systems are studied in two ways - either as one collective system, or as a collection of individuals interacting with each other to produce an overall behavior. The dynamic individual behavior can be studied using mathematical formulas [100] such as differential equations or time-based activities. However, using mathematical equations often restricts models to certain levels of complexity and data being collected. For instance, hierarchical relationships observed at *macro system level*, as well as at *micro internal level* within individuals cannot be easily studied using equations (Figure 1.1). In complex systems, local individual interactions cause emergent system qualities at higher levels, allowing emergence to be a consequence of what happens within these micro levels [114].

TABLE 1.1: Examples of research carried out in complex adaptive systems. Adapted from Schuster [171].

Research Area	Researchers	Year
Darwinian evolution	Smith and Szathmary [189]	1995
Chemical networks	Kauffman [103]	1993
Ecological networks	Sigmund [177]	1993
Insect colonies	Bonabeau and Dorigo [25]	1999
Immune system	Segel and Cohen	2000
Nervous system	Kandel [101]	2000
Economic networks	Lane and Durlauf [11]	1997
Social networks	Frank [67]	1998
Communication networks	Barabasi [4]	2000
Transportation networks	Narguney [137]	2000
Evolutionary games	Hofbauer and Sigmund [85]	1998

1.1 Complex and Adaptive Systems

Multiple disciplines use complex systems to explain unusual phenomena and systems characteristics by artificially creating large simulated systems modeling real systems, aiding understanding on how these systems behave. Table 1.1 discusses some of the early examples in various disciplines and complex system modeling. Some of the common features of these systems are summarized in Figure 1.2.

Individual elements exist on multiple levels within the system, allowing hierarchies, and even hierarchies, to develop where two systems are mutually exclusive and continuously interacting. These elements can act as representatives of either a single performing individual or as a collection of individuals such as groups of multiple individuals. Each element evaluates its behavior based on a reward system and adapts to perform better in the current system conditions. The reward system is determined by a performance measure, where individuals use receptors to read signals and functions to assess these performances.

Adaptiveness of elements is a unique feature that complex systems possess. Researchers have studied how systems predict and readjust efficiently to changing conditions. For example, Hopfield [91] showed that system adaptiveness is highly influenced by presence of noise and attractors in the system. Attractors are environmental points that cause elements to deviate from their ideal paths of behavior, possibly when systems start to show chaotic behaviors.

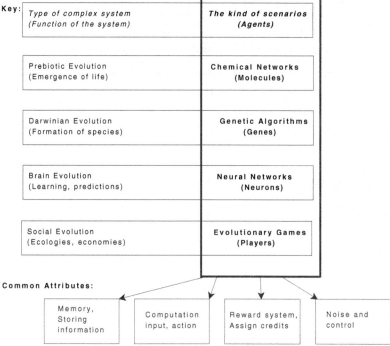

FIGURE 1.2: Examples of complex adaptive systems, their models and common characteristics. cf. [171].

1.2   What Is Chaos?

In mathematics, chaos theory is the description of a dynamic system that exhibits high sensitivity to initial conditions of the system. Conversely, chaotic behavior, in common language, also translates into an unpredictable or unperceived behavior. *Chaos*, thus, has multiple meanings depending in the context it is used. In this book, a chaotic effect refers to an emergent behavior which is unpredictable, or otherwise unknown to observer at the beginning of the simulation. There is a separate research field which involves measuring chaotic points or attractors in a system during simulations, usually measuring initial conditions and then comparing them to a series of outputs generated. Testing these effects of chaos theory is out of the scope of discussions presented here.

Complex systems are known to sometimes go into chaos. Derived from ancient Greek [145], it describes a state that lacks order or even predictability. Langton [114] coined the term 'edge of chaos', which was used to describe the point at which system starts exhibiting chaotic behavior, or the point at which it becomes extremely sensitive to initial conditions. This sensitivity sometimes produces bifurcations (or branches into two possible behaviors) that are difficult to predict (Figure 1.3).

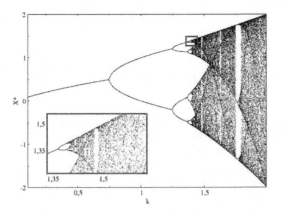

FIGURE 1.3: Bifurcation diagram in a logistic map. Adapted from [122].

1.3   Constructing Artificial Systems

Complex systems can be seen as large problems that can be solved as collections of smaller problems. For instance, ant colonies and individual ant behavior are being studied to give possible solutions to computer networking problems [170], or understanding how prices behave in stock markets.

Large engineering applications, also made up of tiny parts working together, can have precise predictable behavior. These individual units always perform as they ought to, unless they fail due to some dependencies which were difficult to predict. Economic systems also exhibit a wide variety of emergent behaviors, with humans sometimes not paying their credit bills regularly or buying houses without paying mortgages, cited as some of the reasons for 2008 credit crunch [33]. This unpredictability and randomness of individuals, leads to the failure of large systems performing as predicted. The extent of failure, having a domino effect on surrounding elements, depends on how many individuals deviated, allowing complex systems research to become a multi-dimensional problem with techniques from psychology and behavioral economics, enhanced by methods in computer science.

1.4   Importance of Emergence

Goldstein [74] argues that "emergence refers to rising of novel and coherent structured patterns and properties during the process of self-organization of complex systems". But Anderson [8] points out that due to scale and complexity, it is not necessary that the built model would always turn out to be same as its individual real parts. This notion leads to the fact that emergence itself cannot be defined as a perfect pattern with multiple result possibilities.

> "The ability to reduce everything to simple fundamental laws does not imply the ability to start from those laws and reconstruct the universe. The constructionist hypothesis breaks down when confronted with the twin difficulties of scale and complexity. At each level of complexity entirely new properties appear. Psychology is not applied biology, nor is biology applied chemistry. We can see that the whole becomes not merely more, but very different from sum of its parts." [8]

Modeling a system is the process of creating a replica of the system. This could be done by considering only a few aspects of what is needed to be observed from that system, or what modelers desire to test. For instance, testing small gears working together in a clock could either be tested with individual elements modeled as gears, or whole collection of gears connected to the needle, taken as one individual. Modeling depends on modeler requirements to how they want to represent the system.

The model would also be simulated a number of times to understand its average behavior. Randomness in complex systems can sometimes lead to unpredictable patterns, which makes testing a concrete part of modeling.

1.5   Dynamic Systems

Complex systems can adapt to changing environmental conditions. Their ability to cope with changes and their survival makes systems extremely robust and favorable for inspirations in engineering applications. Traditionally, numerical equations with differentiation are used to represent dynamic systems as functions with respect to time. Examples of such equations are Newton's law of motion for particles and forces, represented as expressions of velocity, acceleration during movement and direction of travel for particles. The Navier-Stokes equations are used to describe motion of fluid substances, used to model behavior of water in pipes. Using these equations can represent the system as a derivative of time [144],

$$\dot{X} = \frac{dx}{dt} = F(x) \tag{1.1}$$

Equation 1.1 shows change in a system represented with time, where $X = (x^{(1)}, x^{(2)}, ..., x^{(k)})$ and k is the number of states of the system. The system state is given as a property (for all elements) as a snapshot at the time. This can include individual element properties, environmental conditions and any other attributes involved. Basically, it is a snapshot of the system at time t, taken between starting time $t = 0$ and time $t = k$. In this way, it is possible to determine how the system looks at time $t = t + 1$, if state at $t = t$ is known.

However, complex systems are emergent systems. This makes it sometimes difficult to predict or anticipate, how the system would look at $t + 1$ as there are too many individual element interactions leading to its snapshot at $t + 1$ due to randomness in individual behavior. These systems are also irreversible, which means it is also difficult to work backwards and deduce what the past state was even if current and future states are known. Researchers can deduce a number of reasons why the system behaved in the way, by running repeated simulations and testing the effect of all elements on overall system behavior.

1.6   Is There Evolution at Work?

Being continually adaptive, systems show continuous dynamic change. This uses fewer or basic starting conditions and assumptions to grow into complex system behavior. As time moves forward, certain conditions can be changed to alter its behavior and future system states. Other components that play a key role include *geography* or locations in the system. Geography can influence in ways such as the following:

Communication span for each individual. Messages or communication between individuals, which are limited to particular individuals in an area. This gives them more information and act accordingly.

Messages influence personal behavior. Received messages can be used to determine the next strategies to play based on the incoming information.

Influence of resource availability. Depending on their locations, each individual has various levels of resource available, that can affect its behavior. For example, in a ant colony model, if an ant comes across a stream of water, it can locally change its on-course path, effectively adapting to the situation and locality. Over time, the system will display a stronger ant path being created, that deviates from water. This is an important ability to ensure survival in changing environments.

1.6.1   Adaptation

Adaptation is the ability of individuals changing their behavior or functions to survive better in their present conditions. Examples include developing ability to run faster or hiding from predators. In evolution, organisms with successful adapting capabilities will grow, improving on their likelihood to survive.

Karl Sims [181] presented his work on artificial life, where he displayed evolution in action by creating a computer simulated block creature that had rectangular blocks hinged together. Each block was flexible and allowed to move, such that the creature could restructure itself to suit to the environment it was in.

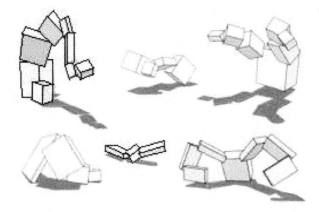

FIGURE 1.4: Examples of Karl Sims's creatures. cf. [180].

The creatures were evolving towards a common goal, which was to swim as fast as they could through a water environment (Figure 1.4). Simulated results showed that there were no optimum solutions, as creatures could not understand water mechanics and its behavior. However, the results showed new designs quickly generated, increasing the rate of survival for creatures in water. Sim's experiments showed evolution was at play when performed with particular goals for survival.

Through various successful adaptations, an emergent behavior can be observed, seen as an outsider view of the system. This is known as *evolutionary drive* in the system, as conditions and time force individuals to change.

Evolution is a term borrowed from biology, where organism populations adapt from one generation to the other. Over time, generations accumulate various differences with each other, depending on how they adapt, allowing new generations to have a gradual divergence from the starting pool of characteristics. These differences are brought about due to locations, limited information or available resources, allowing emergence of new species that can better adapt to given situations. Some individuals not well adapted, will gradually die out, leaving only strong ones to multiply.

The term *reliability* refers to probability of a component operating satisfactorily during a certain time frame. Quantifying reliability requires one to define, assess and combine probabilities of risk and system behaviors [27]. This may require identifying system variabilities and vulnerabilities, to predict lifetimes to assess model reliability.

1.7 Distributing Intelligence?

Evolution is *learning and not intelligence*. Minsky [134] claimed that *intelligence* is used to emphasize swiftness and efficiency of a solution.

> "Evolution's time rate is so slow that we don't see it as intelligent, even though it finally produces wonderful things we ourselves cannot yet make".

Evolutionary behavior can be observed at multiple levels. Every layer can be 'zoomed in' to see different patterns of behavior emerging. Johnson [98] discusses an example of a city as a complex system, where the city itself behaves like one individual system, consisting of a number of thriving neighborhoods within. Each neighborhood consists of a collection of people involved in complex networks such as traffic networks. Similar to ant colonies, a city is a system which has decentralized control, learning from local interactions making a man-made self-organising system using emergence.

However, the beauty of these systems lies in the individual elements. These units can think, restructure and communicate with other units. Some of these characteristics are summarized as follows:

System is part of a larger complex system. The systems are connected to other systems as a hierarchy or using input or output branches.

Systems are open systems. The systems are interacting with other systems continuously with no bounds. In a closed system, the system exists as an isolated entity with specific boundaries, like gas molecules contained in a container, where conditions of thermodynamics hold. Entropy changes can therefore be predicted. However, some systems are a collection of various smaller systems intertwined, have to modeled as open systems.

Systems are dynamically changing. The environment around the individuals is constantly changing, influencing their behavior.

Display emergent behavior. The emerging behavior can be studied at macro levels. For instance, insect colonies achieve their goals quicker by working collusively among individual ants.

Individuals are adaptive. Depending on the changing environment and available system resources, individuals adapt their behavior to survive in given conditions.

Individuals are selfish. All individuals work for their own benefit using local information.

1.8   Modeling and Simulation

Modeling and Simulation (M&S) is a core research area under scientific computing, where artificial systems are created as models and simulated in a virtual environment. Executing them in a virtual environment allows to safely assume changes, in order, to predict how the system would behave when certain changes are introduced in real world situations.

However, it is important to note that a model is only an approximate representation of the system, showing only basic functionalities being explored. It is often a very simple representation of the system, with clearly defined assumptions embed into the model while it is constructed.

A model is a representation of an object, a system or an idea represented in a form other than that of the entity itself [175]. Simulation allows the model to be tested in a virtual world to check its reaction to certain conditions. The model's design would ensure how reliable it is for making predictions.

There are multiple forms in which models are created, such as physical, where models are constructed as prototypes, or scale models, where they represent systems, and mathematical, where models are constructed as analytical mathematical notations, linear and simulation-based representations. In all cases, techniques chosen to construct models, depend on the objectives and aims of the modelers. Model examples include, but are not limited to,

- Engineering applications: Test if certain temperatures will affect smooth running of the engine. These include examples from designing and analyzing manufacturing systems or transport systems.

- Biological models: Of tissues, neurons or cellular models to study effects of chemicals and drugs on cell behavior.

- Economic models: Of various markets such as stock markets, labor markets or economic systems to study the introduction of migration, taxes and money on the overall market behavior.

- Social science models: To study effects of various population dynamics on areas and resources.

- Evaluating systems: Hardware or software performance for a computer system or new military weapons system or tactics on enemy forces.

- Designing communications systems and message protocols for communicating entities.

There are various steps involved in constructing M&S mapping from real world situations and simulating them in a virtual world. The steps involved are as follows:

Step 1. Identify problem being investigated in real world: This is very specific to hypothesis being tested, which can usually not be tested in real or natural settings due to costs or impacts. This justifies it being tried out as a virtual experiment first.

Step 2. Formulate model problem: Formulate a model for a system in a manner by which it is created as a virtual representation. This involves determining assumptions of the model, hypotheses being tested, kind of data being collected from real world to test it and, finally determining which tools to used to create the model. This usually involves talking to domain experts and collecting relevant data to construct most accurate system representations. Computer simulations involve multidisciplinary approaches, where computer scientists work with biologists or economists to construct computational models for systems from their disciplines. A computer scientist has to ensure the model has been correctly represented and all necessary behaviors are captured by it.

Step 3. Simulate model using relevant software toolkits: Use software tools to simulate a model.

Step 4. Analyze data collected: The simulation results are collected and analyzed. The results can be used to find discrepancies and test theory predictions, allowing modelers to verify their models.

Step 5. Data mining techniques: Data analysis techniques such as machine learning, pattern findings and data visualizations help determine the simulation conclusions, in terms of testing the hypothesis.

Step 6. Validity and verification of model: Involve validating and verifying results of simulation, to test if they are correct for conclusions being drawn on hypotheses. At this step, review of the model correctness and result reliability can circle back to step one, by finding issues or wrong assumptions in the initial model constructed.

It is important for developers and researchers to remember that a model is not a goal of the experiment, but it is a process by which simulation will find solution to the hypothesis being tested. Thus the model is only an enabler to the process being investigated [59]. Figure 1.5 shows a flow chart of processes involved when creating biological models. The figure highlights how modelers sometimes need to rework through initial model descriptions, to correct models, after expert advice and results are obtained.

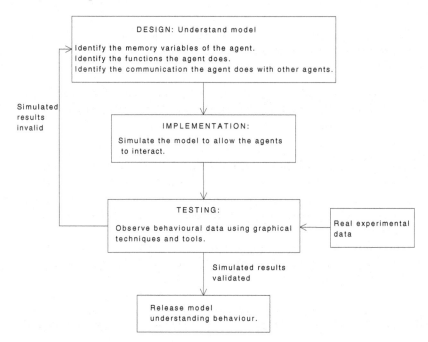

FIGURE 1.5: Modeling process in biology simulations. cf. [107].

1.8.1   Research Examples

Natural Systems

Falling under area of swarm intelligence, ant colonies are extremely efficient in finding shortest possible routes to food in minimum time. Proposed in Dorigo's PhD work [53], ant colony optimization algorithms can solve complex problems like the travelling salesmen problem and network routing problems for dynamic scenarios (Figure 1.6).

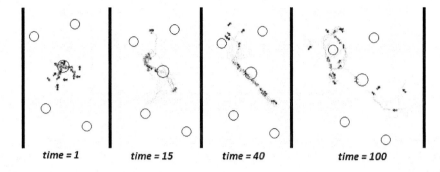

FIGURE 1.6: Various time steps showing ant colonies finding and forming routes to food sources. cf. [23].

Biological systems like the human body are extensively studied as complex systems. The study of NFκB molecule is an example of studying how transcription factors work in cells [149]. Apart from cells, foreign organisms like bacteria, living in human bodies, have also been subject of much research, where bacterial behavior is often studied in human stomachs to determine how they survive in less oxygen levels [125].

Control Engineering

Control systems engineering involves design of robust applications functioning in real world conditions. Research in this area has grown to accommodate various aspects like [95]

- Regulating control of systems.

- Building large systems like bridges or computers.

- Dynamic environmental conditions.

- Optimization and distributing data over large systems.

Being treated as complex systems, systems control theory analyzes large systems as collections of smaller units working together to produce the system output. For instance, Figure 1.7 shows a system made of three interacting units A, B and C. These units can be a capacitor, transistor or a memory chip, working together in the system. The system output produces an effect that brings change in system input at next time step. This becomes an important feedback loop, allowing the system to adapt to changing conditions in the environment (dynamic environments).

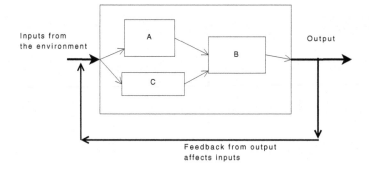

FIGURE 1.7: A system working in the environment. The system is composed of three elements working together to make the system work efficiently. Output produces a feedback, that produces change in the system as time progresses.

Cellular Automata

Cellular automata models have stemmed from basics of computational theory, mathematics and biology. Developed by Ulam and von Neumann [198], they were able to prove the notion of one robot producing another robot or also known as 'the principle of self-replicating systems'. A famous example is the 'Game of Life' by John Conway which uses four simple rules of generations. Here, every element is treated as a cell that transitions based on strict rules predefined by life generations [159].

Being used as a more powerful computational model [203], principles of cellular automata allow individual cells to react and change their states based on their surrounding neighboring cells. If visualized as a plane of cells, there can be a pattern that is observed moving across from one point to other, by subsequent reaction of cells. For example, vibration of molecules in a solid, when provided with heat, acts as a wave propagating from one point to other.

Agent-Based Models

The word *agent* has multiple definitions by different modelers. With respect to agent-based models, the following definition is used in this book:

> "An agent is a computer system that is situated in some environment, and that is capable of autonomous action in this environment in order to meet its design goals" *Wooldridge and Jennings* [205]

It does not specify that every gear in a clock be modeled as one agent or the whole unit to be treated as one agent to reach model goals. This allows modelers to define their own agents and their behaviors per model.

Ideologies surrounding cellular automata models gave birth to concepts of agent-based modeling. Reynolds, in 1985, introduced agent-based models as a driving force for scientific computing, particularly using powerful parallel computers. The computer graphics expert produced the *boids* example which depicted flocking of birds. Later, Langton coined the term *artificial life* to describe similar simulations [201]. These allow simulations of large agent populations to be executed in controlled environments, examining affects of various rules on agent interactions.

Agent-based models encourage bottom-up approaches, allowing research to focus on individual elements interacting with each other, rather than looking at complete scenarios. Initially, pattern in models was proved using differential equations with common examples being found in economic modeling, where mathematical formulas are still being used to prove behavior of ideas. Miller and Page [131] and Epstein [56] have favored agent-based approaches by saying that research should be intensified to focus into agents rather than whole systems, realistically allowing humans to be modeled as agents rather than differential equations.

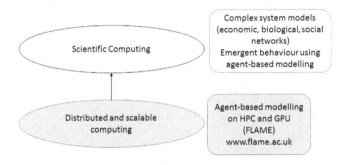

FIGURE 1.8: Research areas of 'Scientific Computing' and 'Distributed Computing' have a close relationship in agent-based modeling.

Advances in parallel and distributed computing can help scientific computation as data and computation grows (Figure 1.8). These can allow data to be processed quickly and analyzed in real time to test models and make better predictions of real complex systems. This work is considerably helped by computing experts in parallel architectures to work with multi-domain scientists to hasten scientific discovery in their fields.

Chapter 2

Artificial Agents

2.1   Intelligent Agents 18
  2.1.1   "Can Machines Think?" 19
2.2   Engineering Self-Organizing Systems 21
  2.2.1   Bring in the Agents 22
  2.2.2   Characteristics of Agent-Based Models 23
2.3   Agent-Based Modeling Frameworks 33
2.4   Adaptive Agent Design 37
2.5   Mathematical Foundations 38
2.6   Objects or Agents? 39
2.7   Influence of Other Research Areas on ABM 40

Engineers and researchers are continually trying to copy or mimic real natural systems to study how they behave. This can allow the design of efficient computer networks or simulate robotic teams to help with disaster recovery in earthquake scenarios. However, studying these real systems is complex and often prone to error and costly.

Modeling and simulation (M&S) allows information to be extracted from real systems, simulate these and test hypothesis without actually practising in the real. Both terms *model* and *simulation* are used interchangeably at times and involve using models, simulators, emulators and toolkits that can write models, execute them and collect their data. Recognized as a separate discipline, M&S has been applied to a range of disciplines such as defense, building and construction, medical sciences and many more.

Figure 2.1 depicts the iterative process of facilitating a scientific method to model observable environments in real world. Data are collected from previously known observations and combined with new methods to create an abstraction of the real world. This scientific method is the process of learning from the real world.

The process of creating a model hypotheses from the real world uses a number of different techniques, like differential equations to represent how system properties change with time, or use Markov models to depict state-based systems. Evolution can be introduced into models to allow learning, such that it evolves into an intelligent system to solve its goals. Minsky [133] has favored use of artificial intelligence techniques to evolve into intelligent machines.

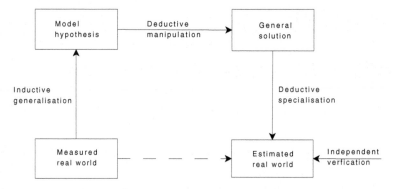

FIGURE 2.1: Scientific method. cf. [61].

2.1 Intelligent Agents

Evolutionary computation is a sub-field under artificial intelligence (AI) research area, involving optimization to automatically solve difficult problems. These contain the following:

- Always include an iterative process where models progressively update their performance.

- Allow growth of given agent populations such that are internally modified based on performance.

- Processes can involve parallel processing.

- Mostly all processes are inspired by principles of natural evolution.

Evolutionary computation contains four sub-topics: genetic algorithms, evolutionary programming, genetic programming and evolutionary strategies. These are shown in detail in Figure 2.2.

Swarm optimization algorithms do not belong to this group, even if used as one of the four approaches. Swarm optimization techniques are inspired from insect colonies and involve large number of individuals working individually to collectively solve the problem. For example, in Figure 1.6, ants could find shortest possible routes to food sources by simply working together and leaving pheromone trails for other ants [37]. These searches are constantly updated depending on food availability and quality.

The focus of evolutionary computation research is mainly the algorithms studying real systems, focusing on optimization and search problems. These problems are difficult to solve and have high complexities, where evolutionary algorithms can keep efficiency high at lower cost.

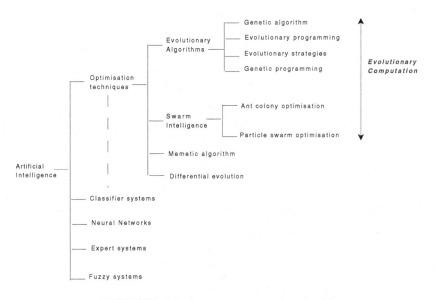

FIGURE 2.2: Separate researches in AI.

Most techniques can be used together such as the swarm intelligence and evolutionary algorithms, each bringing separate characteristics. Some techniques like neural networks and classifier systems can use evolutionary algorithms to improve themselves. Neural networks use evolutionary learning algorithms to allow neural adaptive control in dynamic systems [54]. Classifier systems is a technique which involves using a database of rules and deducing which rules best suit the problems.

2.1.1 "Can Machines Think?"

Posing the question in one of his classic papers, Turing [195] laid the groundwork for AI. He introduced the 'Turing Test', a game which determines if a machine has become as intelligent as a human. The game consists of two players, one being the human interrogator and the second be a machine or another human. The objective is that the interrogator has to determine whether the player being questioned is a human or machine.

Turing machines became the base for defining any kind of computing machine that can solve a given problem. A Turing machine is a machine which reads input symbols of an infinite length tape, processes it, and writes it back to the tape, producing an output. The transition function contains information for machines on what to output and the next position for the tape.

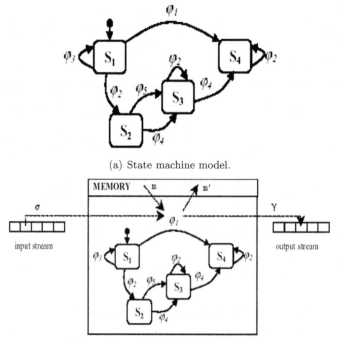

(a) State machine model.

(b) X-machine model with added memory.

FIGURE 2.3: State and X-machine diagrams.

Based on the push-down automata theory, a number of computational methods carry a resemblance to Turing machines. For instance, if there was no input tape, and number of states was finite, such a machine would become a *finite state machine* (Figure 2.3(a)). If the states were added with memory, the machines would then become an *X-machine* (Figure 2.3(b)). Each machine model carries its own properties and varies in computational power by kind of problems it can solve.

Each of the machine models are useful methods by which behavior can be defined. Transition functions, from one state to another, define these complex functions for representing behavior. When self-replicating notions were introduced, Fogel proposed using evolutionary programming techniques to operate on finite state machines to create new finite state machines. Fogel [64, 65] proposed a method by which new machines would '*evolve*' more suited to environment than initial machine configurations. Fogel's work concentrated more on evolution of complete programs, whereas Koza et al. [110] focused on branches within the program to evolve. The following steps can help develop new evolving machines:

1. Create a population of finite state machines.

2. For each machine, observe an input symbol and output produced.

3. Find a method to measure outputs, using a payoff function, known as utility function.

4. Determine fitness of each machine depending on result of the utility function.

5. Machines with a higher payoff are retained to be parents for the next generation of machines. Sometimes half the population is retained until next iterative step.

6. Offspring or new machines can be produced as combinations of two parent machines or by mutation (varying an input symbol or a next state).

Fogel [61] modified complete state machines using evolutionary programs, such as using state machines to play the prisoner dilemma games. These machines were represented as string structures, where genetic operations like crossover and mutation can be applied to modify their structures.

Rechenberg [155] and Schwefel [172] viewed genes as behavioral traits of individuals. Their evolutionary strategies represented the gene as a vector over n dimensions, where mutation and crossover can be performed on n dimensions and on the vector. Holland [88] [89] proposed using genetic algorithm as a search method for adaptive systems. All of these methods use machines to track a particular fitness landscape in a domain to find how far the machine is from ideal. Therefore, this highlights that genetic algorithms in agent architectures would require fitness landscapes to work with.

Put forth as a 'thought experiment', von Neumann [198] presented a hypothetical model of a machine that used raw materials from the environment to produce a second machine by replicating itself. Self-replicating automaton presented the grounds for building cellular automata experiments triggering research in AI, where geography and interactions influenced the production of new machines [105]. Although such a self-replicating robot in the physical world may still be a budding area of research, the concept was introduced and tested in a virtual world of simulation, extended using cellular automata, later giving birth to agent-based modeling methods.

2.2 Engineering Self-Organizing Systems

A model is an approximate representation of a system, showing only basic functionalities or just parts being investigated. Various systems in nature are observed and adopted to create self-organizing systems. Insect colonies, cells and human societies are all examples of these using stigmergy or similar communication mechanisms to form patterns. Stigmergy is a communication mechanism insects use to interact with each other using the environment. For example, ants use pheromones to communicate pathways with other ants. Human societies use messages to communicate information to each other. Similarly, multi-agent systems use communication for coordination in a self-organizing system.

2.2.1   Bring in the Agents

Modeling of complex system behavior is an emergent science which demonstrate complex or social behavior of different communities. Agent-based modeling is a technique which best models these systems, an alternative to conventional differential equation methods. This approach allows a bottom-up procedure, where the focus concentrates on individual interacting units, given clear defined rules and allowed to simulate. The produced emergent pattern of system behavior can then be studied to test and understand behavior of complex systems, otherwise not possible from studying from an outside view.

There are various agent-based environments that can be used to design and test models. Each of these are based on different computational models, varying in computational languages used. Grimm et al. [77] discussed a detailed overview of the problems of verifying models because the tools themselves, are not being designed on predefined software methodologies. They recognize a need for rules to creating agent-based models. Generalizing these rules, allows models to be created with formal methods, encouraging credibility of the results.

Figure 1.5 discussed the process involved in writing an agent-based model. The model starts with a description about individual elements as agents. These agents are using a set of memory variables, functions and communication protocols, that allow them to communicate with each other and the environment. Agents are implemented as separate pieces of code, which communicate using messages.

The individual agent interactions allow certain macro variables to emerge in the system, depicting how whole systems collectively behave. The simulated model can be tested against real data to check its accuracy and validation. However, the complexity in agent-based models increases as,

- Agents can travel in space unlike agents or cells represented in layers of cellular automata.

- Every agent may have limitations in cognitive, physical or temporal abilities based on the model.

- The interaction dynamics between agents lead to emergent patterns to mimic natural system behaviors [174].

- Adopting agent-oriented approaches to natural systems involves modeling many multiple agents. These have heterogeneous structures and are decentralized in nature [207].

2.2.2 Characteristics of Agent-Based Models

Mimicking human societies is a challenge as human behavior varies from person to person, in character and personality. These use various interaction rules that are either defined earlier or introduced during simulations.

Techniques such as genetic algorithms or neural networks can be used to produce agents randomly born with learning or evolutionary capabilities. This means that the agent would act differently if on its own, than when it grows in a group. As Ilachinski [94] argues "emergent properties are properties of 'whole' that are not possessed by any of individual parts making up that whole: an air molecule is not a tornado and a neuron is not conscious".

Figure 2.4 shows a complicated structure of a human agent, modeled by [184]. Different levels of complexity exist, such as sensors, alarms, long-term memory and even an action hierarchy based on particular situations with action priority. Such a model would be increasingly complex in a computational perspective. Making assumptions and specifying model objectives can help abstract some of this complexity, making it easy to model humans in controlled environments.

In terms of intelligent agents, Wooldridge and Jennings [205] have reviewed various techniques for constructing and understanding these. The authors point out that while building intelligent agents one should consider,

Agent theory. Implies that human behavior can be specified as a set of attributes. These attributes can be beliefs, desires or intentions (BDI). A system which has beliefs and desires is a first-order intentional system, whereas a system having beliefs and desires about beliefs and desires is a second-order intentional system. Beliefs are represented as norms in a system like rules in Prolog.

Believe(Mary, world is flat) → Mary believes the world is flat.

These rules are defined as a collection in *possible world semantics* as syntactic representation of languages. It consists of a modal-language (modal operators) and a meta-language (possible world). The latter refers to rule beliefs about goals and correspondence theories [204]. Figure 2.5 describes the various components of strong and weak actions in agents. These use rules to achieve goals and desires. Agent communication languages use KQML (Knowledge query and manipulation language) and KIF (Knowledge interchange format) for message representation.

Agent architecture. These can belong to three strands as follows:

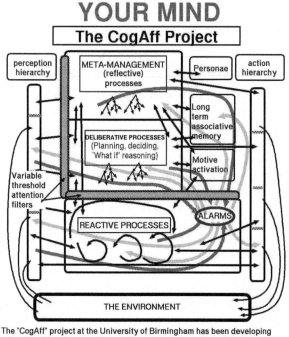

The "CogAff" project at the University of Birmingham has been developing a framework for characterising a wide variety of types of minds, of humans, other animals, and possible future robots. The framework incorporates evolutionarily ancient mechanisms co-existing and co-operating or competing with new mechanisms capable of doing different tasks (e.g. reasoning about what might happen). The figure gives an "impressionistic" overview of some of the complexity. E.g. different sorts of emotions are generated in different levels. More details including papers, slide presentations and software tools can be found at our web site: http://www.cs.bham.ac.uk/research/cogaff and talks directory: http://www.cs.bham.ac.uk/research/cogaff/talks/#talk24
Further information from Aaron Sloman, School of Computer Science

FIGURE 2.4: Your mind designed for CogAff Project. cf. [183].

- Deliberative architecture: Systems can be defined as physical entities of symbols. Deliberative agents use symbols to symbolize scenarios and reason using pattern matching. These agents present two problems:

  - Transduction problem: How to represent real world in symbols.
  - Representation/reasoning problem: Uses symbols to represent real world.

  Planning agents take symbolic representation of world, goals and an action plan to achieve them.

- Reactive architecture: Use of behavior language like symbolic AI. Brooks [32, 31] argued that intelligence is an emergence property of agent interaction and can be viewed from particular perspectives, *"lies in the eye of the beholder"*.

- Hybrid architecture: Combines two features stated above. Agents have representation using deliberative architecture but also a reactive part, such that they react to the environment using symbolic AI.

Additonally, Luck [51] presented more general agent attributes.

- Agent beliefs: Knowledge about itself and environment.
- Agent desires: The states the agent wants to achieve in response to certain actions.
- Agent intentions: Plans adopted by agent.
- Plan library: Agents maintain a repository of available plans.
- Events: Agent actions using beliefs and goals.

Agents may sometimes be required to adopt goals of other agents. This is argues in *Social power theory* where there is a dependence among agents in a network for achieving their own goals. Such a system allows agents to possess resources, creating the divide between some agents being better off than others.

Agent language. Shoham [176] proposed agent-oriented programming as

- Logical system for defining a mental state of an agent.
- Interrelated programming language for programming agents.
- Low-level programs to convert agents in programming language.

Agent languages encompass the implementation aspects and techniques as language representation of agents.

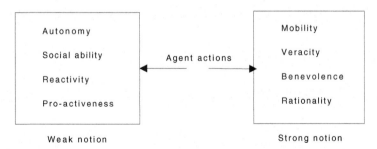

FIGURE 2.5: Weak and strong notions of agent actions. Cf. [205].

Learning in system. Most agent-based systems have mechanisms to learn and adapt their behavior. These agents could be

- Reactive units. Using evolution or Darwinian system, agents react to changes in system and adapt their behavior.

- Goal-directed units. A few agents will be working towards achieving their goals, such as companies taking over other companies for growth or power.

- Planner units. Agents who are goal-directed but also consider environment and goods in their strategy planning.

Agents can use built-in tools to perform parameter learning and assess their actions as they behave in the simulation. This can be achieved through supervised, unsupervised or reinforcement learning, depending on scenarios being modeled. This allows agents to change their strategies or functions depending on personal preferences and information received.

Adaptive agents use multiple methods to learn about the system. Holland and Miller [90] use genetic algorithms to model a population of solutions, coded as strings of characters. Genetic algorithms learn by a biased search towards a combination of solutions, using crossover and mutation. Other methods like classifier systems are an adaptive rule-based system, where each rule is in condition-action (if-then form). The condition allows specific actions to take place.

Reinforcement learning determines how agents can maximize their goals. This differs from supervised learning by finding a balance between exploration and exploitation. The state of the agent at any given time s_t is chosen from a set of allowable states S. The state also determines which action will be chosen $A(s_t)$.

$$s_t \in S \text{ choose action } a \in A(s_t)$$

An agent finds a policy $\pi : S \to A$ to maximize its reward $r = r1+r2+r3$. Various kinds of learning include role learning, learning by discovery and observation through experiments.

Each method is tailored for the problem modeled. Using evolutionary techniques allows agents to make independent decisions because

Agents are autonomous. Agents can operate without intervention of other agents.

Agents are reactive. Agents can read the environment and other agents actions to react accordingly.

Agents are proactive where each agent works to satisfy a specific goal.

Agents are social where they interact with other agents through communication frameworks and alter behavior accordingly.

Reinforcement learning can be used to *teach* agents correct behavior. A simulation game 'Black and White' [191], used reinforcement learning to teach characters the difference between evil and good. The game allowed players to act like gods, with the capability of controlling creatures. Alternately, evolutionary programs in games can also be *taught* to allow new state machines to modify or evolve to generate more intelligent programs (e.g. in a game Rougelike [50]). To achieve this, a reward structure is used in conjunction with evolutionary algorithms to modify behavior, with reward acting as a payoff (fitness).

However, reinforcement learning, in itself, is very limited as it focuses only on agent performance. Agents can use it to choose different behaviors and modify a set of allowable actions to optimize behavior, adapting at t.

Other research in evolutionary concepts in computer science are summarized below,

Turing [195]. Recognized the connection between evolution and machine learning.

Friedman [69]. Proved *thinking machine* can be used in playing chess games.

Friedberg [68]. Improved search space for good programs with given possible solutions.

Bremermann [30]. Presented a multi-objective solution to a numerous parameters in a function. "to a stable point ... [which] need not even be a saddle point [of the fitness function]."

Reed [158]. Used evolutionary algorithms in poker games. Presented use of crossover to find quicker solutions.

Minsky [133]. Objected to Friedberg's solutions saying that they take too much of time to compute.

Fogel [62] [63]. Combined finite state machine with payoff function for producing evolving machines.

Fogel and Burgin [66]. Introduced evolutionary concepts for gaming.

Rechenberg and Schwefel [156] [173]. Produced evolutionary strategies.

Holland [88] [89]. Worked on genetic algorithms for adaptive system.

Bäck and Schwefel [15]. Compared results of experiments for varying crossover and mutation rates.

Turing [195] showed how evolution can aid machine learning by generating new state machines through trial and error. While, Friedman [69] coined term 'thinking machines', using mutation and selection methods in evolutionary processes to give birth to new machines. These efforts introduced the criteria of a 'gene' in evolutionary terms that can be modified in computations. Genes can be defined in various ways such as single alleles, like the ATGC in a human gene with four alleles. For example, Figure 2.6 depicts a computer program represented as a tree structure and a string vector.

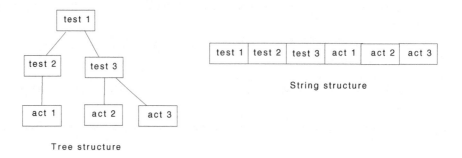

Tree structure

FIGURE 2.6: Program represented as a tree and a string. cf. [50].

Another view is of Mayr's [129], where the author describes evolution as an optimization process, where through learning the system gets progressively better. However, evolution involves alot of trial and error, with new generations having better chances of survival in new conditions.

Synchronization and memory. Gilbert and Terna [73] represented object-oriented languages with efficient memory management and time scheduling to model agents. As stated "with such high-level tools, events are treated as objects, scheduling them in time-sensitive widgets (such as action-groups)." Objects can be tagged with time stamps for execution.

Different agent-based modeling frameworks handle synchronization problems differently. For example, SWARM updates its environment every time an agent does something. While, FLAME waits until the end of an iteration to update changes.

Event-driven versus time-driven. Simulation can either be an event-driven or time-driven. The event-driven approach allows a time step to be updated after any event is triggered. Event-driven agent-based model is a model where changes in state of the system is defined by certain events. For example, an agent becomes active or inactive, depending on memory variables, denoting a progression in the system.

A time-driven system is determined by specific time lengths, which contain a number of actions performed within a time frame. An agent is required to finish all actions during that time step for the system to move forward.

Swarm is an example modeling framework that works as an event-based model. In time-driven approach, the system is updated at end of a function map. FLAME works on a time-driven approach with synchronous updating where all agents are updated at same time and in parallel. Asynchronous updates in a model take place when agents are updated in random order or based on their own internal clocks.

Distributed. Multi-agent systems are concerned with distributed and coordinated problem solving. Bond and Gasser [26] describe distributed AI in three areas:

- Distributed problem solving (DPS): how a problem is divided among a number of nodes to be solved in parallel using knowledge sharing.

- Multi-agent systems (MAS): concerned with 'coordinating intelligent behavior among a collection of autonomous intelligent agents'.

- Parallel AI (PAI): concerned with performance like different computational speeds and finding new paths for problem solving.

Some agent-based modeling frameworks use CNET protocol, which work on principle of a manager managing a set of workers. Every task is decomposed into smaller subtasks and suitable nodes are selected to work on the sub-task. At the end, the results are then integrated together for a complete solution.

Decentralized behavior. Complex systems are decentralized and individuals make decisions based on their locations. Each agent evolves depending on information received locally. Whereas, evolution is based on private memory and messages. Over time niches form, where some agents do better than others.

Messaging. This is an important aspect of agent-based models allowing communication between agents. These interactions are responsible for emergent behavior. This follows the distributed nature of agent-based models, where messaging ensures all messages are read before decisions are made.

Parallelism in agents. Some agent production systems use if-then statements to update rules. These rules determine the next state moved to. A knowledge database is plugged into resolve and execute the rules. Various parallel AI languages, like Prolog, can be used to code these examples. However, it is important to parallelize work and synchronize among all agents to share information. Example factors considered with parallelism are

- Task parallelism
- Match parallelism

- OR parallelism
- AND parallelism

Agents are basically separate modules of code, heterogeneous in nature, but sometimes similar in activities. They need to communicate and prevent agents from accessing same resources leading to deadlocks in the system. Some examples of parallelizing algorithms commonly used are [166],

Algorithms which inhibit dependency. Firing of one rule deletes or adds new rules to the database. Output dependency causes new rules to be added to the database.

Algorithms which enable dependency. New rules satisfy one of the existing rules.

Divide and conquer. Dividing a problem into sub-problems.

Systolic programming. Parallelism with locality and pipelining based on overlap of communication, mapping of processors is similar to problem for parallelism of Logo-like turtle program. Each process has a position and heading. Activation of programs determines position and heading of new processes.

Lisp small talk. Uses symbolic structures. Lexical scoping and procedural scoping.

Artificial neural networks. Distributed memory, distributed asynchronous control and fault tolerance.

Parallelism in genetic algorithms. Genetic algorithms (GAs) are inherently parallel. The genetic operations of evaluating each strategy to produce new populations with higher average fitness, can be done in parallel. However, Holland's [89] version of genetic algorithms proposed a need for serial execution of code when using crossover between two processes.

Haupt and Haupt [82] discussed that using GAs for tackling complicated engineering problems is computational intensive, but can be made efficient by using the parallel nature of GAs. This results in a speedup of simulations and reduces communication between population 'islands' being evaluated. Islands allows populations to be separated into groups and then evolved separately.

In the case of agents evolving together, they could all select strategies from one pool of a strategy population. This would slow simulations down, as there would be a central agent holding strategies and communicate these to all agents, like using social boards to communicate ideas. To reduce this complexity, agents can be equipped with their own strategy populations of a fixed number of ten strategies, as shown in Figure 2.7. Each agent then evolves using these, similar to memetic algorithms solving an optimization problem.

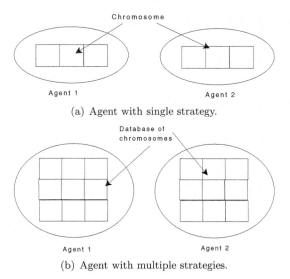

(a) Agent with single strategy.

(b) Agent with multiple strategies.

FIGURE 2.7: An agent can represent a single strategy or multi-strategies.

Agent migration. Agent migration is one of the strongest advantage offered. It allows computation to be extended at a level where space and position are considered, essential in biological and molecular reactions. A few points are,

- Migration reduces much of the network latency as agents perform local interactions independent of complicated network structure.

- Each host should have a platform to incorporate a migrant agent.

- Security issues of agents. Moving agents to a new location could allow access to its internal data easily.

- Agent data should be as minimal as possible to reduce overhead while moving it to a new position.

Modularization and encapsulation. To improve evolvability of programs, they have to be made as independent as possible. For instance, in programming code if-then-else, do-while or for-loops, cannot be fragmented into separate branches. This is because restructuring of code would result in compilation errors. It is essential to make sure the block code does not change its structure.

Modularization is a method which divides the program into functional units. Koza et al. [110] describe a module as a logically closed black box, where only inputs and outputs can be seen, and internal mechanisms are hidden. Each agent can be a module itself or a collection of modules.

Encapsulation is a complete set of program codes as representation of the gene itself. This can be an arithmetic expression represented as string which can be combined with other expressions to find a solution gene.

Automatically defined functions. These ADFs represent a tree structure of a program, and can be of two kinds:

- The result-producing branch evaluated using fitness for that branch.

- Function-defining branch which contains a number of ADFs.

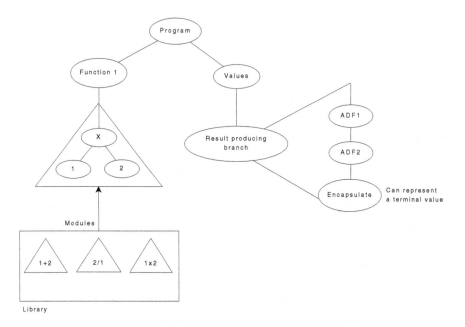

FIGURE 2.8: Evolvability of programs.

Figure 2.8 represents how a program can be represented as a tree structure for evolvability of the program. A program can be broken down into different functions it performs, which can be grouped to form a module. The modules can be stored in a library of modules that hold the genetic makeup of the program. The result-producing branch can be a collection of two ADFs that produce one result that is fed into the program. Koza et al. [110] describe how genetic programming can be used with ADFs.

1. Choose a number of function-defining branches.

2. Fix number of arguments for each ADF.

3. Determine function and terminal sets.

4. Define a fitness measure for each.

2.3 Agent-Based Modeling Frameworks

Over the years, various platforms have been released for agent-based model (ABM) building, each using different programming languages with their own characteristics. Xavier [206] and Railsback [153] provide a detailed comparison between platforms by implementing similar models on them. A comparison of frameworks is shown in Table 2.1.

TABLE 2.1: Comparison of agent-based modeling frameworks.

	SWARM	JADE	MASON	RePast	FLAME
Software methodology	Objective C. Implemented as a nested class structure	Uses FIPA protocols	Java. Implemented as a layered structure	Java	C, XML notations and MPI for parallel computing. Based on X-machine foundation
Easy to use GUI	Yes	Yes	Yes	Yes	Needs integration with other tools
Visualization	3D	3D	3D	2D	2D, 3D
Models executed in Serial/Parallel	Both. Need to wrap Objective C commands in Java for parallelization	Both. Java concurrency commands	Both. Java concurrency commands	Both. Java concurrency commands	Both. MPI libraries for message exchange
Commonly known model examples	Sugarscape, various disciplines	Virus epidemics, Sugarscape	Virus epidemics, Sugarscape, traffic simulation	Mostly social science projects	Skin grafting, economic models

Each of these platform provide modelers with various features. A detailed analysis of frameworks is as follows,

SWARM. This toolkit allows researchers to build agents easily. Built on object-oriented principles, SWARM uses Objective C++ programming language to develop agents as objects with variables and methods. The development involves using inheritance concepts, with new agent classes inheriting from the environment and having their own functions as well. This allows agents to easily communicate with the environment as they inherit variables from the environment.

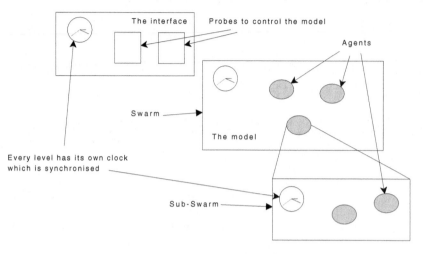

FIGURE 2.9: Nested hierarchy of swarms.

SWARM also supports Java, allowing dynamic functionality by run-time binding with Objective C++, allowing it to run models on parallel machines. Figure 2.9 depicts how SWARM allows nested swarm hierarchies to be developed, with each level to be scheduled with its own scheduler. Agents can be designed to represent other sub-swarms that contain their own set of agents and functions for different timescales. SWARM provides a user-friendly Graphical User Interface (GUI) that allows individual agents to be selected, new attributes to be added and methods to be changed during runtime.

FLAME. Coakley [42, 40] introduced FLAME (Flexible Large-scale Agent-Based Modeling Environment) as an agent-based framework to allow simulations to run on parallel grid architectures.

Formal X-machines were introduced as agent architectures, which allowed mathematical verification of internal agent states by using transition functions. Communicating X-machines were used to communicate using messages as interaction rules as part of agent functions.

FLAME allows deployment of simulations on parallel computers that allow simulations of millions of agents to run in finite time using Message Passing Interface (MPI) libraries for communication messages. MPI is a programming technique used in parallel programming that allows messages from different agents to communicate easily across different processors and platforms. MPI details can be found at www.mcs.anl.gov/mpi/.

MPI allows the model to be independent of hardware platforms involved in the implementation, allowing messages to be packed into certain binders to be sent across nodes easily and independently. A message board is set up from which all agents can read information and perform tasks. To ensure that all functions are in order, there are synchronization points inserted in the model where all data are regulated before moving into the next block of functions.

The synchronization points ensure all memory is synchronized among agents before the agents move into the next block of functions. It is important to synchronize the distributed data so that all agents are aware of information update.

SOAR. Developed in 1983, SOAR is continually being used as an architecture for developing intelligent systems. SOAR has been developed on "the hypothesis that all deliberate goal-oriented behavior can be cast as the selection and application of *operators* to a state" [111]. A state is a representation of the problem-solving situation and goal as desired outcome of the problem.

Based on [112] and [140], intelligence can be functionally described as goals and realizable fundamentals. Newell described that all problems can be broken down into smaller units and solutions of smaller units can be unified for the larger perspective. SOAR allows features like long-term memory, using different memories for different situations and using state hierarchy, where states represent a memory situation. SOAR [84] is being used for military training purposes by the U.S. army, where some work depicts how emotions affect soldier decisions during warfare with emotional agents, making them less predictable to study real situations.

SimAgent. SimAgent toolkit produced at the University of Birmingham was developed as part of the Cognition and Affect project. The project is specifically targeted to design human-like agents and study effects of learning, feeling and emotions. The authors [182] argued that the framework was developed to explore the 'architectural design requirements for intelligent human-like agents'.

"We need a facility for rapidly implementing and testing out different agent architectures, including scenarios where each agent is composed of several different sorts of concurrent interacting sub-systems, in an environment where there are other agents and objects. Some agents should have sensors and effectors, and some should be allowed to communicate with others. Some agents should have hybrid architectures including, for example, symbolic mechanisms communicating with neural nets. We also wanted to be able to use the toolkit for exploring evolutionary processes, as in the 'Blind and Lazy' scenario."

SimAgent has been coded using Pop-11 and Poplog. The programming paradigms [182] use object-oriented programming based on ObjectClass extension to Pop-11. Rule-based programming and pattern matching are also based on Pop-11 with a Pop rule base library. The framework is event-driven, where the toolkit allows events for instance, if the mouse is used to move an obstacle across the scenario, the agent would dynamically calculate their positions and change their walk direction.

Poplog supports Prolog to allow rules and behavior code for logic programming. This allows neural networks to be coded separately and tested [182]. SimAgent also uses an RCLib package for various tests, using neural networks to implement how feelings are handled. Figure 2.4 gives a depiction of how an agent with thinking capabilities is visualized.

Netlogo. Netlogo is a multi-agent programmable modeling environment and is one of the most famous platforms. It allows modelers to give instructions to hundreds or thousands of 'agents' operating independently. This feature makes it possible to explore the connection between the micro-level behavior of individuals and the macro-level patterns that emerge from interactions of many individuals. One of the platform's most efficient feature is its graphical user interface, which provides users with a wide variety of options that can be used to manage models. The GUI is user friendly and very easy to navigate.

Repast. Recursive Porus Agent Simulation Toolkit [7] developed in Java and exploits all of its functionalities. Supported as an open source project, new versions of Repast Symphony can handle high performance computing (HPC) grids and have easy-to-use interfaces for building and modeling agents. The platform can be used to create, run, display and collect data from agent-based simulations and is fully object oriented. Repast is a toolkit with a wide variety of tools and structures. Similar to Netlogo, it also provides an efficient graphical user interface that users can use to manage models, manipulate parameters, set output data and show agent interactions in detail.

JADE. Java Agent DEvelopment framework (JADE) is an agent platform, developed completely in Java and uses Remote Method Invocation (RMI) registry for concurrent connection between machines. Every agent can be defined as a thread, which can simulate as a hierarchy of behaviors. All agents inherit from a class of super agents for common attributes.

JADE is based on standards such as FIPA (Foundation for Intelligent Physical Agents) protocols, used as standard communications languages for agent and environment communication.

MASON. MASON is a multi-agent simulation toolkit that allows discrete events to be simulated. Written in Java, it includes a 2D and 3D library for visualization. MASON has been used to develop ECJ, as a Java-based Evolutionary Computation Research System [123]. ECJ is claimed to be highly flexible with classes dynamically compiled at runtime by a user-provided parameter file.

TAEMS. TAEMS (Task Analysis, Environmental Modeling and Simulation) is described as a 'formal, domain-independent framework' which attempts to solve problems for intelligent agents in different scenarios. The language produces a hierarchical structure of tasks the agent has to perform and assesses them according to goals and deadlines. This hierarchical structure can be viewed as a distributed goal tree, in which branches are joined by AND or OR operations, to produce combinations in scenarios with limited resources and decision-making.

2.4 Adaptive Agent Design

Agents can be designed as either a logical machine with a set of actions or with artificial intelligence, as a set of controllers associated with actions, or with psychology, to mimic minds of real people. However, mimicking the mind of real people is a laborious task and also presents a potential problem to computational complexities of code. Some reasons for this could be the vast amount of memory required, or processing time to pool out relevant information and process it to determine the next action of the agent.

Most researchers have adopted their own methods to achieve the *mind* in their models. Dawid [47] and Vriend [200] have explored use of genetic algorithms for making economic decisions. Sometimes these algorithms can be calibrated to depict decision-making situations, like in works of LeBaron [115] and Marks [127]. Duffy [55] used human subjects as experimental data to calibrate learning in computational agents.

Researchers have often debated that learning architectures in agents can be categorized into specific disciplines - deliberative or reactive [205]. Sloman's [182] work supported use of hybrid architectures, with SimAgent using a detailed agent architecture to encompass most attributes (Figure 2.4). Wooldridge [204] studied developing computational logics behind architectures of multi-agent systems.

2.5   Mathematical Foundations

Multi-agent systems are temporal systems that are highly dependent on time steps. This allows agents to have a set of allowable action set, A_t, made available depending on current time step. Day [49] discussed that these actions chosen by agents at time $t+1$, a_{t+1}, are dependent on various stimuli. These are defined as follows:

$$a_{t+1} = f(o_t, m_t, d_t, x_t, u_t) \tag{2.1}$$

where

- Observation of agent at time $t+1$, $o_{t+1} = \sigma(a_t, s_t)$, where s_t is environment state at time t.

- Memory of agent at time $t+1$, $m_{t+1} = \mu(o_{t+1}, a_t, s_t)$.

- A process of agent at time $t+1$, $d_{t+1} = \pi(m_{t+1}, a_t, s_t)$.

- Plan of agent at time $t+1$, $x_{t+1} = \delta(d_{t+1}, a_t, s_t)$.

- Implementation of agent at time $t+1$, $u_{t+1} = \iota(x_{t+1}, a_t, s_t)$.

The action structure is very explicitly produced by modelers or programmers, as a step-by-step procedure when creating predictable agents. An aspect ignored above is the learning capability of the agent. Most actions may not be chosen during a simulation. The agent should be able to evaluate available actions and modify them to suit its purpose. This process of learning encourages the agent to optimize its behavior, to better suit the conditions at time t. To enable this, the agent code needs a feedback to assess its performance.

Machine learning techniques have used various methods to construct optimizing of artificial agents. Reinforcement learning can allow agents to optimize themselves in dynamic environments. To achieve this, agents have a method to assess their performance in certain situations using a reward structure.

1. At time t, agent sees the environment state, $s_t \in S$ and set of possible actions at this state, $A(s_t)$. Note, previously a set of allowable actions were dependent on time A_t. Now, this is dependent on the environment's state, bringing in awareness of the agent's surroundings, $A(s_t)$.

2. Agent chooses an action to perform, $a \in A(s_t)$.

3. As a consequence of its action, the environment changes its state, $s(t+1)$, and receives a reward or payoff, r_t.

4. Based on inputs, the agent chooses a set of actions to help maximize reward obtained.

Summarizing above, formally a multi-agent system should consist of the following, in addition to Equation 2.1:

- A set of environment states, $s \in S$,

- A set of actions for agent, $a \in A$,

- A set of allowable actions at state s_t for agent $A(s_t) \subseteq A$,

- The action chosen by agent at time t, $a_t \in A(s_t)$,

- A set of scalar rewards r_t received by agent at time t dependent on how it performed at time $t - 1$.

2.6 Objects or Agents?

Code can be objects or agents. The differences are summarized,

- An object is a term that accommodates object-oriented programming principles, which allows objects to relate to other objects, through inheritance and attributes. Agents, however, are complete code pieces that hold all data properties within itself.

- An object allows data size to be reduced by inheriting functions and attributes from parent classes. An agent has a bigger size for an individual, as they contain data and functions with their memory. As agents are isolated and work independently, this is a great advantage in parallel computing, when more than thousands of agents are deployed over processors and minimum communication across processors is preferred. If there is too much communication across processors, this increases computational overhead of messages, introducing latency. All communicating agents can be placed on the same processor to reduce overhead. These are load-balancing issues in parallel computing.

- Agents allow experimenting with bounded information principles.

- Agents can use machine learning techniques by learning. Multi-agent learning can be both cooperative and competitive learning.

"Why can't we build, once and for all, machines that grow and improve themselves by learning from experience? Why can't we simply explain what we want, and then let our machines do experiments or read some books or go to school, the sort of things people do. Our machines today do no such things." [135]

Figure 2.2 depicts different research, stemming from the umbrella of artificial intelligence. Each method is designed for specific purposes, such as genetic algorithms are efficient optimization techniques to search in NP-hard problems or neural networks used to encourage speech and voice recognition in software and other areas. Researchers [167] compared the efficiency of these techniques, when applied to similar problems and drawn conclusions for computational efficiency, resources and time.

Advances in parallel computers and architectures have aided research in multiple areas of science and engineering, with ABM platforms working with researchers with less programming experience. Sante Fe has produced various agent-based models of various kinds like artificial stock market, molecular structures and more. Details can be found at the main website (http://www.santefe.edu/).

2.7   Influence of Other Research Areas on ABM

Markov modeling using Markov decision processes. These models are based on mathematical expressions of Turing machine models. The algorithm involves executing a number of rules, encoded on a symbol string. Markov models can be expressed as chains containing stochastic processes whose states change with time. These state changes carry conditional probabilities associated with them. The future states are independent of past states.

Markov decision processes use a reward function attached with Markov chains. For every transition, the state receives a reward that affects the transition probability of the state. Using reinforcement learning, these systems are useful in dynamic programming problems and training problems such as using unobservable states in hidden Markov models.

Neural networks. These are recreate biological structures of the neuron activity in organisms. The various nodes are connected to each other, with each connection carrying weights for the path to process data. Neural networks can be trained using real data. The simulated data can then be verified if it produces similar results.

Mechanism design (MD). Parkes [146] described mechanism design as a problem for designing a protocol, distributing and implementing particular objectives of self-interested individual agents. An agent makes a decision respecting other agents, based on its own private information and behaves selfishly. The Economics Nobel Prize for 2007 was presented to the Mechanism Design Theory [93]. It follows the "Hayek theory of catallaxy where 'self-organizing system of voluntary co-operation' is brought about as market progress". However, there is criticism to the theory, stating that if MD were used to design markets, some agents still end up monopolizing markets.

Gaussian adaptation. Evolutionary algorithms designed for stochastic adaptive processes take more than one attribute into consideration. The number of samples is denoted by N dimensional vectors to represent multivariate Gaussian distribution.

Learning classifier systems. These LCS use reinforcement learning and genetic algorithms. The rules can be updated using reinforcement learning, allowing different strategies to be chosen.

- Pittsburgh-type LCS - population of separate rule set represented by GA, recombines and produces best of rule sets.
- Michigan-style LCS - focuses on choosing best within a given rule set.

Reinforcement learning. As described above for optimizing behavior.

Self-organizing map. Similar to Kohonen map, it uses unsupervised learning to produce low-dimensional representation of training samples, while keeping the topological properties of input space. Uses a feed forward network structure with weights to choose neurons and produce Gaussian functions.

Memetic algorithms. Learning algorithms, a combination of swarm optimization and genetic algorithms. Each individual program is chosen from a population and allowed to evolve. Each individual uses a learning technique to evolve either Lamarckian or Baldwinian learning. Lamarckain [113] theories use environments to change individuals, known as the adaptive force. Baldwinian [18] uses learning in genetic material of the individual. These are supported by trial-and-error and social learning theories. For instance, trait becomes stronger as a consequence of interaction with the environment. Individuals who learn quickly are at an advantage. Blackmore distinguishes the difference between these two modes of inheritance in the evolution of memes, characterising the Darwinian mode as 'copying the instructions' and the Lamarckian as 'copying the product' [24]. Each program is treated as a meme. The next step involves these memes to coevolve to fit the problem domain.

Chapter 3

Designing X-Agents Using FLAME

3.1   FLAME and Its X-Machine Methodology 44
    3.1.1   Transition Functions 47
    3.1.2   Memory and States 47
3.2   Using Agile Methods to Design Agents 48
    3.2.1   Extension to Extreme Programming 51
3.3   Overview: FLAME Version 1.0 51
3.4   Libmboard (FLAME message board library) 54
    3.4.1   Compiling and Installing Libmboard 55
    3.4.2   FLAME's Synchronization Points 57
3.5   FLAME's Missing Functionality 58

The Flexible Large-scale Agent Modeling Environment (FLAME) was developed through collaboration of the Computer Science Department at University of Sheffield (UK) and the Software Engineering Group (STFC) at Rutherford Appleton Laboratory, Didcot (UK).

The framework is a program generator that enables creation of agent-based model simulations that can easily be ported onto high performance computing grids (HPCs). The modeler defines models using XML notations and associated code for agent functions is given in C language. FLAME is able to use its own templates, to generate serial or parallel code automatically, allowing complex parallel simulations to execute on available grid machines.

FLAME agent models are based upon extended finite state machines (or X-machines) that allow complex state machines to be designed and validated. The tool is being used by modelers from nearly all disciplines - economics, biology or social sciences to easily write their own agent-based models, run on parallel computers, without any hindrance to the modelers to learn how parallel computing works. The toolkit was released as an open source project, in 2010, via its web page (www.flame.ac.uk).

Agent structure, their messages and functions are defined in the model description file. The model description file, written in an XML format, is fed into the FLAME framework to generate a simulation program. The simulation program generator for FLAME is called the Xparser (Figure 3.1), which is a series of compilation files, compiled with GCC (Gnu Compilers) and accompanying files, to produce a simulation package for running simulations. Various parallel platforms like SCARF, HAPU or IceBerg have been used

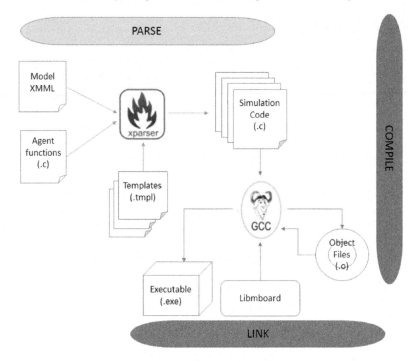

FIGURE 3.1: Block diagram of FLAME. cf. [76].

in the development process to test the efficiency of the FLAME framework
[41, 39].

3.1   FLAME and Its X-Machine Methodology

FLAME agents are based on mathematical notation of formal X-machines,
to represent the agent architecture, their memory, messages, states and transition functions. Compared to traditional state machine models representing
an agent, extended state machines (X-machines) are a powerful model that
can represent complete definitions of agents. With added complexity, memory
and communication protocols, communicating X-machines can easily be used
to mathematically define and verify large complex systems as they interact
through messages.

An X-machine can be formally stated as [104],

$$\boldsymbol{X} = (I, O, M, S, F, T, IS, IMS) \qquad (3.1)$$

where

- I is sequence of inputs to machine $\mathbf{X}$,

- O is sequence of outputs of machine $\mathbf{X}$,

- M is memory of machine $\mathbf{X}$,

- S is sequence of states of machine $\mathbf{X}$,

- F is set of functions $(F : I \times M \longrightarrow O \times M)$ of machine $\mathbf{X}$,

- T is set of transitions $(T : S \times F \longrightarrow S)$ of machine $\mathbf{X}$,

- IS is machine's initial state $\mathbf{X}$,

- IMS is machine's initial memory state $\mathbf{X}$.

Figure 3.2 describes a state and a corresponding X-machine state diagram, for an ant which forages for food and travels back to the nest. Figure 3.2(a) shows the state machine diagram having more number of states and transition functions as comapred to the X-machine in Figure 3.2(b). The X-machine can represent most of the complexity by functions acting to its memory, not possible as a state model. The transitions between states are a result of these functions and not conditions (which is seen in state machines).

The transition functions are also dependent on memory of the ant, being updated whenever there is a change in state. The memory can contain information such as variables to 'stay in nest' or 'move' in 'Moving Freely' (Figure 3.2(b)).

Every state in an X-machine diagram shows the state of the memory. For example, when the ant is 'at-nest', in the memory this is represented as the nest coordinates. In this state, the ant can perform only certain functions such as *staying-at-nest, move, move to food* or *ignore food*. Depending on these functions, the ant can change its memory state, allowing another set of functions to become available to the ant. It can decide to 'look for food', *lift food* or *get lost* in the surroundings.

X-machines can represent more detailed agent descriptions, memory and functions, more suitable to design computational agents and also based on mathematical foundations. A basic definition of an agent $\mathbb{A}$ is

1. A finite set of internal states

2. A set of transition functions operating between states

3. An internal memory set which is finite

4. A language for sending and receiving messages between agents

$$\mathbb{A} = (\Sigma, \Gamma, \mathbb{Q}, \mathbb{M}, \Phi, \mathbb{F}, q_0, m_0) \tag{3.2}$$

where

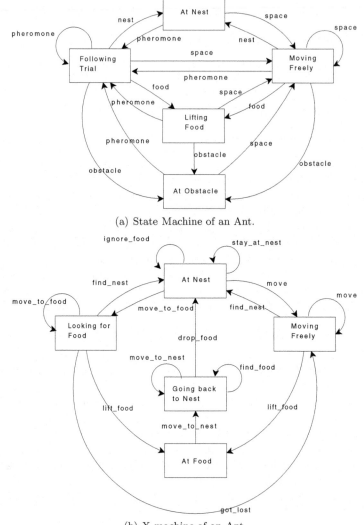

(a) State Machine of an Ant.

(b) X-machine of an Ant.

FIGURE 3.2: State and X-machine diagrams of an ant foraging for food. cf. [104].

- Σ are a set of input alphabets,

- Γ are a set of output alphabets,

- $\mathbb{Q}$ denotes set of states,

- $\mathbb{M}$ denotes variables in memory,

- Φ denotes set of partial functions, that map input and memory variables to output and a change on memory variable. The set $\Phi : \Sigma \times \mathbb{M} \to \Gamma \times \mathbb{M}$,

- $\mathbb{F}$ is transition function to next state, $\mathbb{F} : \mathbb{Q} \times \Phi \to \mathbb{Q}$,

- q_0 is initial state and

- m_0 is initial memory of the machine.

3.1.1   Transition Functions

Transition functions allow agents to change their state to modify their behavior. These require inputs on their current state $s_{(1)}$, current memory values m_1, and the possible arrival of a message at time t_1. Depending on these three variables, the agent changes its state to another s_2, updates its memory to m_2 and optionally sends another message t_2. Some transition functions may only perform a function on the memory, where messages are empty $\emptyset$, or with some data.

$$Message = \{\emptyset, < data >\} \qquad (3.3)$$

Agent transition functions are expressed as a set of stochastic rules with time.

3.1.2   Memory and States

The differences between internal states and internal memory sets allow a flexibility in modeling systems. There are situations where agents have only one internal state and various complex variables defined in memory. Equivalently, agents can have simple memory variables, but a large state space with multiple memory functions.

Software behavior has traditionally used finite state machines to model a system as inputs and outputs. More abstract system descriptions have included UML (Unified Modeling Language) notations [205], but these are mainly diagrammatic representation, lacking writing and testing simulation code descriptions.

Testing a system, specified as a finite state machine, allows its behavior, expressed as a graph, for traversals of all possible and impossible executions of the system. Testing an X-machine, with memory, follows main stages.

- Identify system functions.

- Identify states which impose order of function execution. For each state, identify the memory as a set of variables, accessed by outgoing and incoming transition functions, similar to the process of branch traversal used in testing methodologies.

- Identify input and output messages.

As shown in Figure 3.3, FLAME agent architectures contain,

- A finite set of internal states of agent,

- Set of transition functions that operate between states,

- An internal memory set of the agent,

- A language for sending and receiving messages,

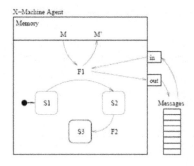

FIGURE 3.3: FLAME uses strict X-machine architecture - Memory, Functions, States and Messages.

3.2   Using Agile Methods to Design Agents

Agile software development encourages principles for developing software, where requirements and solutions evolve through collaboration between clients and developers. Extensively being used in industrial software engineering practices, the approach has proven very successful by promoting collaboration, adaptive planning, early delivery and continuous improvement for the product being delivered. Figure 3.4 presents a mapping of how agile methods incorporate agent-based modeling development as a software project.

Incremental software development or agile methods were a reaction to

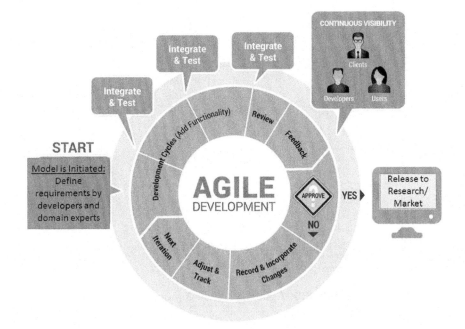

FIGURE 3.4: Incorporating agile methodology in agent models. Modified from [20].

traditional, rigid, micro-managed software deliveries, where customers were often not happy by the software products received. Agile methods introduced new ideas with various enhancements to product development cycles using scrum, extreme programming (XP), adaptive software development or even feature-driven development. Based on the Agile Manifesto [20], it follows,

1. Customer satisfaction is achieved early via continuous delivery of the product.

2. Remain flexible to change requirements.

3. Working software is delivered every few weeks.

4. Nearly everyday cooperation is between developers and clients.

5. Projects are built via motivated individuals and teams trust each other.

6. Encourage face-to-face meetings.

7. Development is maintained at a constant pace.

8. Attention is given to technical design and excellence for product.

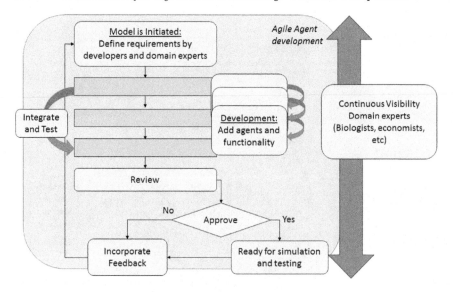

FIGURE 3.5: Agile agent development process.

9. For simplicity, try not to increase work if not needed.

10. Allow products to emerge from teams.

11. Teams continuously reflect their performance, share ideas and learn, becoming effective.

Developing agent models, using agile allows multiple domain experts to work together, to develop software models. It enables computer scientists to work closely with domain experts, to build a model based on domain requirements. Testing of model and verification is also continuously done at every stage of the release, minimizing risk of wrong assumptions being implemented in the model.

Agent-based models are difficult to implement due to sheer complexity of models. Through the process in Figure 3.5, domain experts can interact closely with modelers, to monitor model development and research hypotheses. However, with these advantages, the process sometimes slows down development and introduces the need for continuous client involvement. But at the end of every cycle, as the model matures, the clients are able to monitor and develop ideas and test these through their models before releasing them to the research domain.

3.2.1   Extension to Extreme Programming

The classic approaches to developing software include following the waterfall or spiral models of development. In general, these include stages of requirements analysis, developing specifications, design and architecture, coding, testing, documentation and maintenance. However with their advantages, there were a number of issues leading to the development of using scrum and XP approaches.

Agile methodologies involve multiple interactions and software evolves through different phases. Extreme programming also falls under Agile software development methodologies and stresses customer satisfaction. It empowers programmers to respond to the customer changing demands, emphasizing team work by giving equal opportunities. Eventually teams become productive and self-organize for efficiently problem solving. It uses five essential building blocks:

- Communication: within team and with customer.

- Simplicity: Change requirements as per customer needs and deliver early.

- Feedback: Testing starts from day one.

- Respect: within team and customer.

- Courage: to rebuild if necessary.

Small and functional releases of code are done regularly, where customers can evaluate and have visibility at all times. Given the nature of building agent-based models, XP is an ideal process of developing these. It recognizes that all requirements will not be known at the beginning of the model and may change as it develops. The team can plan small releases, accommodating tools to build communication and continuous development improving design.

Developers can use X-machines and XP approaches together, to help develop list of inputs, processing functions, outputs and encapsulate these as agents. These can then be extended to test-driven approaches for testing correct agent behavior with FLAME.

3.3   Overview: FLAME Version 1.0

FLAME can model various levels of complexity - from modeling molecules to modeling complete human communities. FLAME does this by only changing agent definitions and functions [42]. An agent architecture with characteristics is shown in Figure 3.6 (Figure 3.3).

- The simulation contains multiple types and agent concentrations, that are of similar kind or behave differently across scenarios.

Structure of an Agent

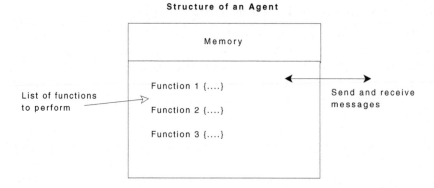

FIGURE 3.6: Structure of basic agent. Agents represent any individual such as a household, an ant or a firm.

- Agent memories enable heterogeneity in agent population, representing unique qualities.

- Agent performs a list of functions as *actions* in a scenario, depending on model design.

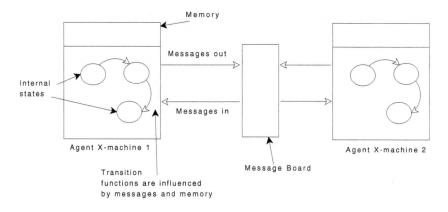

FIGURE 3.7: Two X-machine agents communicating through a message board. The message board library (Libmboard) saves current active messages during the simulation time step.

The X-machine agents communicate through messages, using interaction rules specified in model description (XML) files. These involve posting to and reading from message boards, shown in Figure 3.7. FLAME thus follows these steps:

- Identify agents and their functions.

- Identify states which impose some order of function execution within agents.

- Identify input and output messages of each function (including possible filters on inputs).

- Identify memory as a set of variables, accessed by functions (including possible conditions on variables for functions to occur).

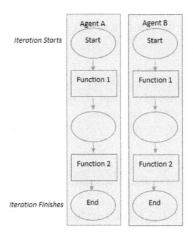

FIGURE 3.8: One iteration with two agents, each with two functions.

Figure 3.8 shows a basic two-agent structure, with two functions, without any interaction between them. The agents have a start state and traverse states, until they reach end state. This process runs during one time step or 'iteration'. Figure 3.9 shows how transition functions perform on agent memory, reading and writing to it.

FLAME provides a number of advantages for writing agent-based models:

- Ease of programming using C language.

- Ease of parallel computing, enabling the possibility of having a large number of agents on parallel processors.

- The agent architecture defined using X-machines. This architecture adds flexibility as additional memory and functions.

- Because the back-end is written in C language, it is easy to allow the framework to communicate with other languages if required.

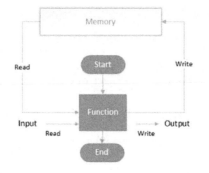

FIGURE 3.9: Transition functions perform on memory variables.

- Further operations such as graphics and genetic operation libraries can easily be merged into the framework, or adapted with FLAME's own libraries added as code extensions.

3.4   Libmboard (FLAME message board library)

Messaging results in the communication overhead when agents send and receive messages. The Message Board Library, designed by STFC, was built to handle these communications in an efficient manner. Agents post their messages to local message boards, where all agents can read, instead of sending messages individually to agents. These message boards are regularly synchronized to prevent irregular data repetitions.

FLAME uses message passing interface (MPI) to allow platform independence in implementation, by allowing the message to be packed into a certain binder, which can be sent across nodes easily and independently. A message board is set up from which all agents can read information and according to their functions perform tasks. The library uses distributed memory model Single Program Multiple Data (SPMD) paradigm to communicate messages efficiently (Figure 3.11).

Various experiments were performed to measure how messaging time can affect simulation time of an experiment [41]. Table 3.1 discusses how simulation times change with number of processors and platforms for the same EURACE model (Figure 3.13) [39]. The experiments vary simulation times by varying the message filters, agent partitioning in geometric or round robin arrangements and various HPC architectures. Varying the agent distribution across the nodes can affect how many agents have to communicate across

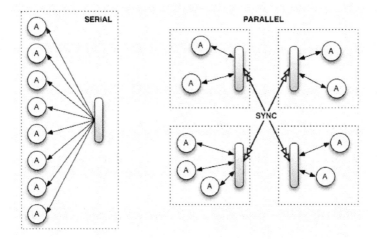

FIGURE 3.10: Serial versus parallel execution of agents.

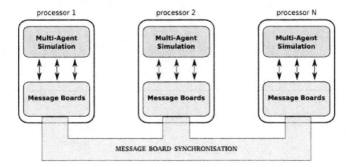

FIGURE 3.11: Distributed memory and synchronization.

nodes and message filters are embed in XML notations to reduce number of messages being searched through by agents (Figure 3.12).

3.4.1   Compiling and Installing Libmboard

The version of Libmboard used with FLAME version 1.0 was "libmbord-0.2.1" built using a Linux instance. Provided with two folders 'src' and 'build' for compiling, it is linked with FLAME executable files. After building, the next command extracts the zipped Libmboard folder to an existing folder. After successfully extracting Libmboard, enter the local Libmboard folder to build and install it.

```
> mkdir ~/src  ~/build
```

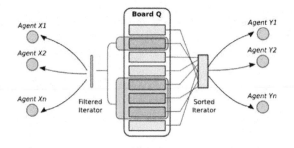

FIGURE 3.12: Using filters and iterators to quicken message parsing for agents.

TABLE 3.1:   Simulation times across multiple processors in HPC grids [76].

Number of Processors	HAPU	NW-Grid	Hector
2	92.3	43.2	-
4	43.3	32.4	29.8
6	76.9	29.3	26.2
8	63.6	30.8	24.1
10	72.1	37.3	22.9
12	34.6	36.5	22.0
14	82.5	40.5	22.1
16	45.0	41.0	21.7

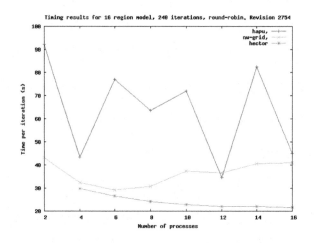

FIGURE 3.13: Simulation times across multiple processor nodes.

```
> /dev/null
> tar zxvf libmboard-0.2.1.tar.gz C ~/src
```

```
> cd ~/src/libmbord-0.2.1
> ./configure disable-tests with-mpi=/usr/lib64/mpich
--prefix=HOME/build/libmboard
> make
> make install
```

3.4.2   FLAME's Synchronization Points

FLAME produces a state dependency graph for each model that contains information on function order executing in one iteration. Common parallelization problems occur due to occurrence of deadlocks during execution [43]. A deadlock occurs when

- A resource is not in mutual exclusion condition, where the resource cannot be used by more than one process at one time.

- Processes which are holding resources wait for more new resources.

- No resource can be forced to be removed from the process using it, until released by the process.

- Two or more processes form a circular condition, where one process is waiting for the second process to release a resource, at the same time when the second is waiting for the first to finish working with it.

FLAME agents communicate through messages being written and read via the message board library (Figure 3.11). Using a model description file, it works out possible synchronization points between functions in both serial and parallel nodes (Figure 3.10). These points create a function interaction dependency with the message board library, making sure all information is homogenized for all agents. This ensures that deadlocks can be prevented when the model runs on parallel computers.

Figure 3.14 shows how synchronization points are set between messages being sent and read by functions. At a synchronization point, the message board for a message list is locked for reading. This allows all agents to send messages to the message board, before any other agents can start to read them.

This approach is good for parallel computing, but also prevents agents from executing any dependent functions until all messages are read. All functions that involve reading that message board are then scheduled to run after the function sending the messages has finished. This allows all agents to follow particular plans and cannot change their behavior during a time step.

If agents need to change their functions, this is done by implementing a rule database or adding flags to which function to execute. The synchronization points have to be scheduled around these choice functions. FLAME can also specify a message range to build local social circles and message filters to enable neighborhood emergence.

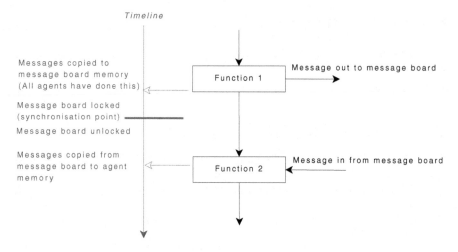

FIGURE 3.14: Timeline showing when the synchronization point occurs when messages interact with functions.

3.5   FLAME's Missing Functionality

Using the X-machine approach provides agents with much needed complexity to model and simulate complex models. However, there are a number of advantages other frameworks provide, which FLAME currently does not. This makes it necessary for modelers to add more complex code, embedding complex behavior into agents.

Static global conditions. Agents behave in a world with no changing conditions, as there is no global environment agent acting as the world. None of their decisions have any effect 'on the world'. It only acts as a space. This requires a central agent to be programmed into the system if the model needs a world representation.

No learning or adaptation in agent functions. Agents cannot learn about their performance and adapt to new conditions. This would require additional programming to add a reward function and complex function choices to show adaptation.

Assumption of perfect rationality in agents. Agents have access to complete message boards and perfect knowledge, unless randomness is added to the message choice.

- Modelers assume agents have perfect knowledge of the past and the present, including the model they exist in.

- Agents are also assumed to have perfect foresight.

- All agents are homogeneous or heterogeneous, represented as different values for the same memory variables.

- All agents have a maximum expected utility in decision-making.

The lack of networks. Most agent communications are based on networks formed [75]. FLAME models are independent agents with no links between them. FLAME does not create neighborhoods, unless specified.

There are additional restrictions in FLAME, because of its parallelization ability and synchronization points in agent functions and messages. These points ensure that all prior agent functions are finished, before the simulation progresses. At these points, the message board is locked, until all agents have finished sending their messages to that particular message board. The board is only unlocked when it is read by agents. Figure 3.14 depicts this along with a time line to show when these occur.

Synchronization points are useful to remove deadlocks, but makes agents wait to finish processing before it can continue. However, this architecture also makes the system extremely predictable. The agents have pre-knowledge of what to do and makes it impossible for them to learn and change their behavior. This prevents emergence to occur at lower levels of these systems, only seen above by predictable behavior below. In order to encourage emergence, as in real systems, modelers need to overcome these assumptions:

- Adaptability of agents is ignored.

- Cannot allow agents to think.

- Allow emergence in the model at lower levels.

- Somehow not to compromise on the parallelism effort of the models, else it will crash during execution.

Chapter 4

Getting Started with FLAME

4.1   Setting Up FLAME 62
    4.1.1   MinGW 63
    Configuring MinGW for Windows users: 63
    4.1.2   GDB GNU Debugger 63
    4.1.3   Dotty as an Extra Installation 64
4.2   Messaging Library: Libmboard 64
4.3   How to Run a Model? 65
4.4   Implementation Details 65
4.5   Using Grids 68
4.6   Integrating with More Libraries 69
4.7   Writing a Model - Fox and Rabbit Predator Model 71
    4.7.1   Adding Complexity to Models 72
    4.7.2   XML Model Description File 72
    4.7.3   C Function 76
    4.7.4   Additional Files 81
    4.7.5   0.xml File 83
4.8   Enhancing the Environment 84
    4.8.1   Constant Variables 84
    4.8.2   Time Rules 84

FLAME stands out from other agent-based modeling frameworks, allowing the parallel deployment of the simulations on large parallel computers using *Message Passing Interface* (MPI). MPI is used to send messages between agents, located on different nodes or processors on various platforms. This capability allows FLAME to run large simulations (up to 500,000 agents) in a matter of minutes, enhancing research in time and complexity in the written models [107]. FLAME reads the model files and automatically generates a large simulation program in C which can be run in parallel by using the '-p' flag or in serial by default.

The input files defined by the modeler are

- Model.xml - Multiple xml files containing the whole description of the model such as agent definitions, memory variables, functions and messages between them.

- Functions.c - Multiple '.c' files contain implementations of agent functions, names of which are specified in the xml files.

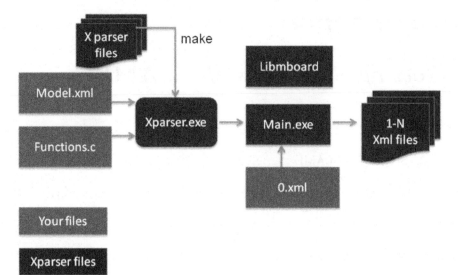

FIGURE 4.1: Block diagram of the Xparser, the FLAME simulation component. Blocks in blue are files automatically generated. The green blocks are modeler's files.

- 0.xml - This contains initial states of memory variables of agents, initialization of all memory parameters.

The number of the resulting XML files depends on number of iterations specified in the model run (through Main.exe).Figure 4.1 shows which parts of the model are written by modelers and which are the code produced by Xparser. Figure 4.2 shows how the software blocks exists and interact with the input files in the block diagram.

4.1   Setting Up FLAME

FLAME executes on Windows, MacOS and Linux operating systems. The list of files required:

- Latest version of the framework (Xparser)

- C compiler such as the GNU compiler

- Libmboard files for parallelization tailored for all windows, macos and linux users.

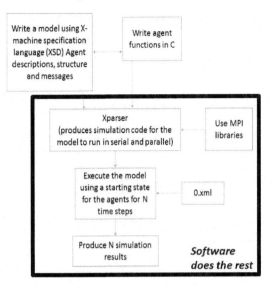

FIGURE 4.2: FLAME software blocks.

4.1.1 MinGW

The recommended C compiler is MinGW, built in Unix. Windows users can download a copy from the following link: `http://sourceforge.net/project/showfiles.php?group_id=2435&package_id=240780`

Configuring MinGW for Windows users:

Computer → Properties → Advanced settings → Environment variables → System variables

Select 'Path' and edit it as follows:

- Add the path of the MinGW after ';'

  `C:\MinGW\bin`

- Rename

  `C:\MinGW\bin\mingw32-make.exe`

  to 'make.exe'

4.1.2 GDB GNU Debugger

For debugging, GDB GNU Debugger is recommended. It is freely available with how-to-use tutorials. You can get your free copy from the following link:

http://sourceforge.net/project/showfiles.php?group_id=
2435&package_id=20507

Note: Windows users are recommended to use the version 5.2.1 (available at the above link).

4.1.3   Dotty as an Extra Installation

The parser creates diagrams about the function flow in the model. These are created in an '.dot' format. In order to view these files, download Graphviz from www.graphviz.org. Grpahviz also allows the files to be converted into other forms of images or formats like pdf.

4.2   Messaging Library: Libmboard

The Libmboard files help with parallelization of the models. These read the messages during the simulation and create message boards, to efficiently manage messages. Working on various platforms, versions of Libmboard have been created to allow the execution either on Windows, Linux or MacOS platforms.

For Windows: Download Libmboard for Windows. Unzip and place the folder where the model is. This is an already precompiled version for Windows platforms.

For Linux/Mac systems:

1. Download the latest version of Libmboard from ccpforge. Place this anywhere, and point to it when running the simulation.

2. Go to the folder where Libmboard is to be placed. For example, in 'Volumes', use following command to make a directory for Libmboard:

   ```
   > mkdir libmboard
   ```

3. Unzip and access the downloaded Libmboard folder.

   ```
   >./configure --prefix =/Volumes/libmboard --disable tests
   > make
   ```

   Once successful, move to folder 'Volumes/libmboard'

   ```
   > make install
   ```

   This will compile the Libmboard on the system.

4.3 How to Run a Model?

Go to the Xparser folder and compile it, 'make' (MinGW).

```
> cd xparser
> make
```

This compiles the Xparser and generates 'xparser.exe'. The simulation program then uses the model XML description to parse the model.

```
> xparser ..\model\model.xml
```

The model can then be compiled to generate simulation files and create the 'main.exe'.

```
> cd ..\model
> make
```

For Mac/Linux users:
For compiling the simulation program on Mac/Linux users, the Libmboard folder needs to be documented with the Xparser global file. This is done by compiling the model with specifics on the Libmboard location.

```
> make LIBMBOARD_DIR= /Volumes/libmboard
```

The model executes for number of iterations, using the 0.xml file as initial memory.

```
> main.exe 100 xml\0.xml
```

This command runs the model for 100 iterations, generating 100 XML files.

4.4 Implementation Details

The FLAME Xparser generates compiling files for the model when simulated. Taking inputs - the model XML file, template files and agent function files the Xparser generates the following files:

- Doxyfile - Generates project-related configuration options such as output folders, input files specification and path builders.

- Header files for every agent with '.h' prefix - Generated with a model xml file as input, with pointers to access the agent memory variables.

- Header.h - Generates specifications for agents as xmachine data structures and transition functions. Also generates message board-related iterators.

- Low_primes.h - Defines arrays holding the prime numbers to aid with partitioning.

- Main.c - Holds main functions of the program, how the simulation reads data, parses and produces simulation code.

- Main.exe - Executable file for the simulation.

- Main.o - Object file.

- Makefile - Contains details of file paths and information needed for executing code.

- Memory.c - Holds memory functions on how to read, access and write to agent memory. Also handles the memory functions of message boards. Functions allow looping through messages, free message boards (MB_Clear()), initialize pointers to every state in agents, create message boards (MB_Create()), get and set functions.

- Memory.o - Object file.

- Messageboards.c - Holds functions for message board - creating, reading, iterating and deleting message boards when needed. Example adding a message:

```
MB_AddMessage(b_messagename, &msg)
```

Reading a message:

```
Inline static message * getInternalMessage(void)
rewind iterator MB_Iterator_Rewind(i_message_to_use);
```

And accessing messages:

```
MB_Iterator_GetMessage(i_message_to_use, (void **)&msg);
```

- Messageboards.o - Object file.

- '.o' files for each agent functions file - Object files.

- Partitioning.c - Helps with partitioning of the data, with functions on geometric and round robin partitioning. Also contains functions on how to save data to local nodes when simulation is distributed over multiple nodes. The file does partitioning, cloud data array initialization, temporary node creation for adding agents. It also creates in-tags to reference agents and reads its memory values, and eventually creates the machines.

- Partitioning.o - Object file.

- Rules.c - Functions created to handle conditions in the model. The file has been deprecated in FLAME v1.0. Functions can contain conditions on timings such as

```
iteration_loop%20 ==6 return 1 else return 0;
```

- Rules.o - Object file.

- Timing.c - Holds functions to handle calling on time to read in calender time or the day.

- Xml.c - Contains functions on reading and writing to an XML file. In C language, this is handled by reading the data held by each tag and parsing through them. Same process while writing an xml file. The functions allow reading of specific data variables like static arrays of int, float and more, along with writing data structures, defined in the model xml file.

- Xml.o - Object file.

The templates help generate the above files dynamically when the xparser compiles the model. The parser then reads the model file, and generates the dependency graph (stategraph of the model) and above files. The graph describes the layers in which the agent functions will execute and creates state flow diagrams of the agents. The Make file creates all necessary files that are linked for execution.

While simulating, the parser reads the 0.xml file in 'r' mode, checks partition method as geometric or round robin and reads the values from 0.xml to assign agent memory. If partitioning, the agents are allocated according to the position x and y across the grid (SPINF stands for extreme values of grid). As the agents are read, they are added to a linked list structure, where the first state of the agent is used as an argument such as

```
add_Person_agent_internal(current_Person_agent, Person_00_state);
```

The memory values are copied into the agent memory, using the in_tags defined earlier, and functions created for writing values such as

```
write_int_static_array(*file, temp, size)
```

During the simulation, the calculated agent values are written to the next iteration xml files, produced and saved with the last states of the agents.

The main.c file initializes the message board environment (MB_Env_Init()). This involves initializing pointers, iteration numbers, calling read initial states and generating partitions (cloud_data, total nodes, partition_method). The simulated data are saved in an iteration xml file and write log files. If all functions are traversed successfully, at the end the x-machine structures are freed.

Synchronization is handled by the sync code such as

`MB_SyncStart(messagename) and MB_SyncComplete(b_messagename)`

The iterator for the message board is created and randomized as follows:

`MB_Iterator_Create(b_messangename,&i_messagename)`

In the end, the iterators are deleted (MB_Iterator_Delete(messagename)) and agents are freed to transit to the next state. At the end, all message boards are cleared (MB_Clear(b_messagename)), write data in files, move agents to their start states and clean up. It is important to randomize iterators here, to ensure that agents reading the data will eventually not be biased towards the first few messages in the board, especially if agents make decisions on the first few messages received.

Further details on the message board can be found at `http://www.softeng.cse.clrc.ac.uk/wiki/EURACE/MessageBoards/mplementationNotes/API`, where messages can be customized to allow quicker processing and reduce overhead in simulations. These customizations are of three types:

1. Immediate messages: Messages are read and then deleted. Examples such as a rabbit is eaten or job filled, where message is read, perform functions and then delete it.

2. Counter messages: Messages which exist for some time, such as cost or price message, read by all agents.

3. Handshake: Messages generate new message by reading an old message, such as a hired message, read by one agent in response to a previous message.

Customizing messages to ensure quicker processing are subject to further research in parallel computing, to see how models can be made more efficient over various machines.

4.5   Using Grids

FLAME was compiled to be executed on the High Performance Computing (HPC) grid Iceberg (`http://shef.ac.uk/wrgrid`). The steps involved were

1. Connect to Iceberg grid via a username and password using a terminal window.

2. Copy files to Iceberg which include xparser and libmboard files.

3. Copy files from Iceberg which include the resulting xml files.

4. Configure the C++ compiler and MPI libraries on Iceberg. This was done by creating a symbolic link to all required files like cc1plus.

5. Configure Libmboard by running ./configure. The MPI libraries are linked here.

6. Run a model on Iceberg in parallel by using the following commands:

```
./xparser path_of_model.xml -p
```

For example,

```
./xparser ../model/turningKernel.xml -p
make CC=/usr/local/packages5/openmpi-gnu/bin/mpicc
LIBMBOARD_DIR=/home/ac1mk/libmboard export
LD_LIBRARY_PATH=\$LD_LIBRARY_PATH:/usr/local/packages5/
openmpi-gnu/lib
```

Run on parallel nodes by submitting a job.sh file which contains the number of nodes and the time you specify to it.

```
#!/bin/sh
#\$ -l h_cpu=0:30:00 (Specifies how long the job is likely
to last}
#$ -cwd
#$ -pe openmpi-ib 8 (Specifies the openmpi-ib
library should be used, with 8 cores)
#\$ -q parallel.q (Specifies the queue to be used)
#\$ -v SGE_HOME=/usr/local/sge6_2 (Specifies the path)
/usr/mpi/gcc/openmpi-1.2.8/bin/mpirun
/home/ac1mk/trial2/main 100 trial2/output/p8/t100k/0.xml -r
```

On a Mac system this could be done directly by

```
mpirun -np numberOfNodes ./main numberOfIterations 0.xml -r
```

```
mpirun -np 16 ./main 100 0.xml -r
```

4.6 Integrating with More Libraries

FLAME uses C, XML and interacts with generated files. The models can be enhanced by using the same principle and working with additional libraries. For example,

- More C standard libraries and custom libraries: C Math functions, C memory accessible variables (include the basic int, char, float) or all C functions. Custom libraries are user-defined libraries.

- MPICH-2 libraries: Implementation of MPI, MPICH2 provides MPI implementations for important platforms and massively parallel processors. It is open-source and freely available for use in parallel programming environments. OpenMP can be used with MPI to allow hybrid parallelization for loop-level parallelism. More information on integrating MPICH with Windows and Linux platforms can be found at [78].

- OpenGL libraries: Open Graphics Library provides access to functions for high quality graphical image in 2D or 3D. OpenGL is concerned with manipulation of frame buffer for drawing and rendering of images. It can be integrated with C language for its functionality.

- Libxml2: With XML input and storage format, Libxml2 is an XML C parser toolkit that can be used across various platforms. It provides a variety of language bindings and wrappers making it useful with various languages. It provides support for Document Object Model as well.

- Interfacing with SBML: Libsbml allows manipulation of various SBML (systems biology markup language) files and data streams. Written in C and C++, it is used as a library for various programming languages (like C/C++, Java, Lisp, Perl, Matlab) and makes the code portable to different platforms of Windows or Linux.

- HDF5: Hierarchical Data Format 5 is a library used to store various data. It can allow data to be stored as dataset or in groups. A dataset is a multidimensional array of data elements whereas a group is a structure for organizing objects. Using these two storage mechanisms, one can generate any kind of required data structure - like images, arrays of vectors or grid structures.

- GraphViz graph library: FLAME is already using GraphViz for generation of dotty diagrams or graphs showing function dependencies in parallel activity. It can be used for more outputs on networking structures, depicting hierarchy, clusters and more.

- Sqlite3: A small C library supports the SQL database engine to store data into a single disk file. These files can be shared as a database between various machines.

4.7   Writing a Model - Fox and Rabbit Predator Model

Simple Fox and Rabbit Model Scenario: "Foxes are chasing the rabbits, and rabbits are moving around randomly in a 2D scene".

The above scenario describes a basic predator model, where foxes are chasing rabbits to consume them. The model writing steps include

1. Identify agents in the scenario: Fox Agent and Rabbit Agent.

2. Identify memory of each agent, based on the scenario: (Fox Memory - x position, y position, Fox agent id), (Rabbit Memory - x position, y position, Rabbit agent id).

3. Identify functions of agents in the scenario: (Fox function - Chase rabbit), (Rabbit function - Move randomly).

4. Identify messages being communicated between various agents: (Rabbit location message).

5. Using the information above, draw a block representation of how the agent would perform during one iteration of the simulation (Figure 4.3).

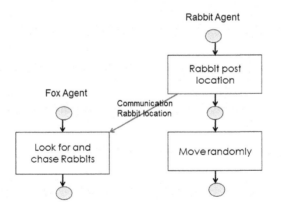

FIGURE 4.3: Flow diagram for the simulation describing agents, its functions and communications.

4.7.1 Adding Complexity to Models

Once an initial specification of the model has been drawn up, modelers can update the description with more complex behavior for agents (Figure 4.4). Table 4.1 shows the model details such as agent memory variables, functions and messages generated.

Updated Scenario: "Foxes are chasing the rabbits, and rabbits are dodging the foxes. The foxes have a life expectancy of 10 days. Assuming every iteration is representing a day."

TABLE 4.1: Model parameters for fox and rabbit example.

Model Definition	Variable
Agents	Fox and Rabbit
Fox memory	x position, y position, fox agent id, life Expectancy
Rabbit memory	x position, y position, rabbit agent id
Fox functions	Chase rabbits, check life expectancy
Rabbit functions	Dodge foxes
Messages for fox agent	Output: fox location message; Input: rabbit location message; Output: eaten message
Messages for rabbit agent	Output: rabbit location message; Input: fox location message; Input: eaten message

4.7.2 XML Model Description File

Models descriptions are represented in XML formats, allowing them to be human and computer readable. The DTD (Document Type Definition) of the XML document for FLAME has gone through various updates (located at http://www.flame.ac.uk/docs/), modifying and adding xml tags as the models became more complex in FLAME simulations. In general, the model file needs to contain the basic elements:

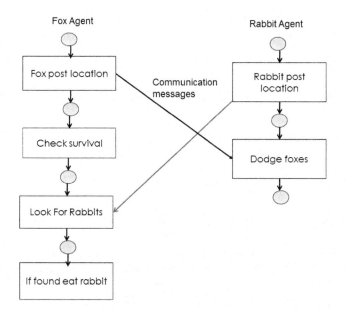

FIGURE 4.4: Flow diagram for simulation describing agents, their functions and communications between the agents with complexity.

- Other model files (either enabled or disabled),
- Environment,
- Constant variables,
- Function files or links to them,
- Time units to represent the period of frequency during the simulation,
- Data types or structures,
- Agents with name, description, memory, functions,
- Messages communicated in the model: name, description, variables.

```
<?xml version="1.0" encoding="ISO-8859-1"?>
<xmodel version="2">
<name>Predator Model</name>
<author>Authors Name</author>
<date>190207</date>

<!--*** Environment values and functions ***-->
<environment>
  <functionFiles>
```

```
      <file>functions.c</file>
    </functionFiles>
  </environment>

  <!--**** X-machine Agent - Fox ********-->
  <agents>
    <xagent>
      <name>Fox</name>
      <!-- Variables -->
      <!-- All variables used by Fox are declared here
           to allocate them in memory -->
      <memory>
        <variable>
          <type>int</type><name>foxID</name>
          <description></description>
        </variable>
        <variable><type>int</type><name>lifeExpectancy</name></variable>
        <variable><type>double</type><name>foxX</name></variable>
        <variable><type>double</type><name>foxY</name></variable>
      </memory>

      <functions>
        <function>
          <name>foxInformation</name>
          <description>send location message</description>
          <currentState>00</currentState>
          <nextState>01</nextState>
          <outputs>
            <output><messageName>foxInformation</messageName></output>
          </outputs>
        </function>

        <function>
          <name>foxSurvives</name>
          <description>check fox life</description>
          <currentState>01</currentState>
          <nextState>02</nextState>
        </function>

        <function>
          <name>chaseRabbits</name>
          <description>find rabbits</description>
          <currentState>02</currentState>
          <nextState>03</nextState>
          <inputs>
            <input><messageName>rabbitInformation</messageName></input>
          </inputs>
          <outputs>
            <output><messageName>rabbitEaten</messageName></output>
          </outputs>
```

```
          </function>
       </functions>
   </xagent>
<!--*** End of Agent - Fox ***-->

<!--*** X-machine Agent - Rabbit ***-->
   <xagent>
   <name>Rabbit</name>
   <!-- Variables for the Rabbit -->
   <memory>
      <variable>
        <type>int</type><name>rabbitID</name>
        <description></description>
      </variable>
      <variable><type>double</type><name>rabbitX</name></variable>
      <variable><type>double</type><name>rabbitY</name></variable>
   </memory>

   <functions>
      <function>
        <name>rabbitInformation</name>
        <currentState>00</currentState>
        <nextState>01</nextState>
        <outputs>
          <output><messageName>rabbitInformation</messageName></output>
        </outputs>
      </function>

      <function>
        <name>dodgeFoxes</name>
        <currentState>01</currentState>
        <nextState>02</nextState>
        <inputs>
          <input><messageName>foxInformation</messageName></input>
          <input><messageName>rabbitEaten</messageName></input>
        </inputs>
      </function>
      </functions>
   </xagent>
   <!--*** End of Agent - Rabbit ***-->
</agents>

<!--*** Messages being posted by the agents to communicate ***-->
<messages>
   <!--  Message posted by foxes -->
   <message>
      <name>foxInformation</name>
      <description>Fox location message</description>
      <variables>
        <variable><type>double</type><name>foxX</name></variable>
```

```
          <variable>
            <type>double</type><name>foxY</name>
            <description></description>
          </variable>
        </variables>
      </message>

    <message>
      <name>rabbitEaten</name>
      <description>Rabbit eaten message</description>
      <variables>
        <variable><type>int</type><name>rabbitID</name></variable>
      </variables>
    </message>
    <!-- Message posted by rabbits  -->
    <message>
      <name>rabbitInformation</name>
      <description>Rabbit information message</description>
      <variables>
        <variable><type>int</type><name>rabbitID</name></variable>
        <variable><type>double</type><name>rabbitX</name></variable>
        <variable><type>double</type><name>rabbitY</name></variable>
      </variables>
    </message>
</messages>
<!--**** End of Messages ********-->
</xmodel>
```

4.7.3   C Function

Function files define the source code for implementing agent functions. These are included in the compilation script (Make file) of the produced model.

```
<functionFiles>
  <file>agent_1_source.c</file>
  <file>agent_2_source.c</file>
</functionFiles>
```

Modelers can access agent memory variables by using CAPITALS such as FOXX, FOXY for agent memory variables as defined in the xml file. This file needs to be included in the functions files by the following command:

```
<agentname>_agent_header.h
```

A few rules when writing agent functions involve:

- All agent functions should return '0'. If the function returns '1', the agent dies in the simulation.

- Agents can create static or dynamic arrays in agent memory or make some locally, to manipulate within the function.

- Messages can be manipulated, such as by:

    - Add message with:
      Add_messagename_message(var1,var2)

    - Loop through messages with:
      START_MESSAGENAME_MESSAGE_LOOP
      Messagename_message->variables
      FINISH_MESSAGENAME_MESSAGE_LOOP

For example:

```
while(rabbitInformation_message)
  {
    /* Access data from message */
    rabbit_id_found = rabbitInformation_message->rabbit_id;
    if(rabbit_id_found==1)
    {
      printf("Rabbit with ID =1 is found!")
    }
    /* Traverse through next message */
    rabbitInformation_message = get_next_rabbitInformation_message
                    (rabbitInformation_message);
  }


/* Example Fox functions */
/** \fn Check if fox survives malnutrition
*/
int foxSurvives()
{
  /* For each time step lower life expectancy */
  LIFEEXPECTANCY = LIFEEXPECTANCY - 1;
  /* Check if dead */
  if(LIFEEXPECTANCY == 0)
  {
    printf("Fox dies of hunger\n");
    /* Kill dead fox agent */
    return 1;
  }
  return 0;
}


/** \fn Fox_location()
 * \brief Send message with fox location
 */
int foxInformation()
{
```

```
  /* Send fox location message */
  add_foxInformation_message(FOXX, FOXY);
  return 0;
}


/** \fn Chase_rabbits()
 * \brief Read rabbit locations and chase
 */
int chaseRabbits()
{
  /* Closest rabbit id, default for no rabbits is -1 */
  int closest_rabbit_id = -1;
  /* Shortest distance, used to find closest rabbit */
  double shortest_distance = 9999.0;
  /* Current distance squared, holds
  distance from current agent to message sending agent  */
  double current_distance_squared;
  /* Holds position of closest rabbit */
  double closest_x, closest_y;
  /* Angle to closest rabbit */
  double theta;

  /* Look for nearest rabbit */
  /* Get first rabbit location message from list */
  rabbitInformation_message = get_first_rabbitInformation_message();
  /* Loop through all messages on the list */
  while(rabbitInformation_message)
  {
    /* Calculate distance */
    current_distance_squared =
                (rabbitInformation_message->rabbitX - FOXX)*
                (rabbitInformation_message->rabbitX - FOXX) +
                (rabbitInformation_message->rabbitY - FOXY)*
                (rabbitInformation_message->rabbitY - FOXY);

    /* If distance within view distance of the fox */
    if(current_distance_squared <=  (fox_view_length*fox_view_length))
    {
      /* If shortest distance then save values */
      if(current_distance_squared < shortest_distance)
      {
        shortest_distance = current_distance_squared;
        closest_rabbit_id = rabbitInformation_message->rabbitID;
        closest_x = rabbitInformation_message->rabbitX;
        closest_y = rabbitInformation_message->rabbitY;
      }
    }
    /* Get next message */
    rabbitInformation_message = get_next_rabbitInformation_message
```

```
                               (rabbitInformation_message);
  }
  /* If there is a rabbit close */
  if(closest_rabbit_id != -1)
  {
    /* Is the closest rabbit in eating distance */
    if(shortest_distance <= (fox_eat_length*fox_eat_length))
    {
      printf("Eat rabbit %d\n", closest_rabbit_id);
      /* Send eaten message to rabbit */
      add_rabbitEaten_message(closest_rabbit_id);
      /* Move to rabbit position */
      FOXX = closest_x;
      FOXY = closest_y;
      /* Increase life expectancy */
      LIFEEXPECTANCY = LIFEEXPECTANCY + 10;
    }
    else /* Else chase closest rabbit */
    {
      /* Calculate angle to rabbit */
      theta = atan((closest_y - FOXY)/(closest_x - FOXX));
      /* Move run length of fox towards rabbit */
      FOXX = FOXX + (fox_run_length * cos(theta));
      FOXY = FOXY + (fox_run_length * sin(theta));
    }
  }
  else
  {
    /* Move randomly */
    FOXX = FOXX + (fox_run_length - (rand()/(double)(RAND_MAX)
                    *(fox_run_length*2.0)));
    FOXY = FOXY + (fox_run_length - (rand()/(double)(RAND_MAX)
                    *(fox_run_length*2.0)));
  }

  /* Mirror location off boundary */
  FOXX = handle_boundaryX(FOXX);
  FOXY = handle_boundaryY(FOXY);
  return 0;
}
/* Example Rabbit functions */
/** \fn Send message with rabbit location
*/
int rabbitInformation()
{
  /* Send rabbit location message */
  add_rabbitInformation_message(RABBITID, RABBITX, RABBITY);
  return 0;
}
```

```
/** \fn Dodge_foxes()
 * \brief Read fox locations and dodge
 */
int dodgeFoxes()
{
  /* Use idea of force from foxes to move rabbits */
  double fox_x_force = RABBITX;
  double fox_y_force = RABBITY;
  /* Angle to move rabbit, calculated from fox force */
  double theta;
  /* Distance from current agent to agent sending message */
  double current_distance_squared;
  /* Fox count */
  int foxes = 0;

  /* Check if eaten, by reading eaten messages */
  rabbitEaten_message = get_first_rabbitEaten_message();
  while(rabbitEaten_message)
  {
    /* If message relates to me then die */
    if(rabbitEaten_message->rabbitID == RABBITID)
    {
      printf("Rabbit %d dies\n", RABBITID);
      return 1;
    }
    rabbitEaten_message = get_next_rabbitEaten_message
                (rabbitEaten_message);
  }
  /* Dodge foxes, by reading fox location messages */
  foxInformation_message = get_first_foxInformation_message();
  while(foxInformation_message)
  {
    current_distance_squared =
                (foxInformation_message->foxX - RABBITX)*
                (foxInformation_message->foxX - RABBITX) +
                (foxInformation_message->foxY - RABBITY)*
                (foxInformation_message->foxY - RABBITY);

    /* If distance within rabbit view distance */
    if(current_distance_squared
      <= (rabbit_view_length*rabbit_view_length))
    {
      /* Add fox location to fox force */
      fox_x_force += foxInformation_message->foxX;
      fox_y_force += foxInformation_message->foxY;
      /* Increment fox count */
      foxes++;
    }
```

```
          foxInformation_message = get_next_foxInformation_message
                        (foxInformation_message);
  }
  /* If foxes in view distance */
  if(foxes)
  {
    /* Use fox force to calculate angle to move */
    theta = atan((fox_y_force - RABBITY)/(fox_x_force - RABBITX));
    /* Move rabbit run distance along angle */
    RABBITX = RABBITX + (rabbit_run_length * cos(theta));
    RABBITY = RABBITY + (rabbit_run_length * sin(theta));
  }
  else
  {
    /* Else move randomly */
    RABBITX = RABBITX + (rabbit_run_length -
            (rand()/(double)(RAND_MAX)*(rabbit_run_length*2.0)));
    RABBITY = RABBITY + (rabbit_run_length -
            (rand()/(double)(RAND_MAX)*(rabbit_run_length*2.0)));
  }

  /* Mirror location off boundary */
  RABBITX = handle_boundaryX(RABBITX);
  RABBITY = handle_boundaryY(RABBITY);
  return 0;
}
```

4.7.4 Additional Files

Additional files can be accompanied by model files. These allow modelers to organize their code and easily manage complex functions. For instance,

- Modelers can list global values in an additional header file. This allows modelers to change these values before runtime and maintain one file.

- Modelers can add additional functions used by agents in a separate functions file, included at runtime.

```
/**Example Library header file**/
#include "header.h"
#include "Fox_agent_header.h"
#include "Rabbit_agent_header.h"

/** \def Distance foxes can see rabbits */
#define fox_view_length 20.0

/** \def Distance foxes can eat rabbits */
```

```
#define fox_eat_length 1.0

/** \def Distance foxes can run */
#define fox_run_length 1.5

/** \def Distance rabbits can run */
#define rabbit_run_length 1.0

/** \def Distance rabbits can see foxes */
#define rabbit_view_length 10.0

/**Example Library functions file**/

/** \fn Handle agent positions with respect to the boundary
 * \param position The current position in one axis
 * \return The new position mirrored along the boundary
 */
double handle_boundaryX(double position)
{
  double newPosition = position;
  if(position < 0.0) {
    newPosition = (-1) * position;
    position = newPosition;
  }
  if (position < 20.0) {
    newPosition = position + 20;
    position = newPosition;
  }
  if(position > 950.0) {
    newPosition = position - (position - 950) + 20;
  }
  return newPosition;
}


double handle_boundaryY(double position)
{
  double newPosition = position;
  if(position < 0.0) {
    newPosition = (-1) * position;
    position = newPosition;
  }
  if (position < 20.0) {
    newPosition = position + 20;
    position = newPosition;
  }
```

```
  if(position > 600.0) {
    newPosition = position - (position - 390) + 20;
  }
  return newPosition;
}
```

4.7.5   0.xml File

The 0.xml file represents the starting state of the model, which is the starting memory value of agents at initialization stage. The example shows that the model starts with two foxes and two rabbits. These values are updated as the simulation progresses, producing more iteration files. These successive iteration files contain updated memory values of agents mainly their X and Y positions (Figure 4.5).

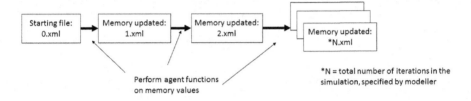

FIGURE 4.5: Iteration files with updated agent memory results.

```
<!-- Example 0.xml file-->
<states>
<itno>0</itno>
<xagent>
  <name>Fox</name>
  <lifeExpectancy>20</lifeExpectancy>
  <foxX>29.113054</foxX>
  <foxY>8.329377</foxY>
</xagent>
<xagent>
  <name>Fox</name>
  <lifeExpectancy>20</lifeExpectancy>
  <foxX>83.390059</foxX>
  <foxY>29.294528</foxY>
</xagent>
<xagent>
  <name>Rabbit</name>
  <rabbitID>1</rabbitID>
  <rabbitX>65.316155</rabbitX>
```

```
  <rabbitY>13.165019</rabbitY>
</xagent>
<xagent>
  <name>Rabbit</name>
  <rabbitID>2</rabbitID>
  <rabbitX>0.208702</rabbitX>
  <rabbitY>79.911496</rabbitY>
</xagent>
</states>
```

4.8   Enhancing the Environment

The environment tag in the XML file hosts additional tags for information, which may be required by the parser for efficient simulation of the model. Following are tags that can be defined.

4.8.1   Constant Variables

Constant variables refer to global values used in the model. These can be defined in separate header files which can then be included in one of the functions. The header would look as follows:

```
**
 * \file  my library header.h
 * \brief Header for user created library functions. */

#define fox_location_msg 100.0
#define application_msg 100.0
#define time_msg 100.0
#define message_range 100.0

void bubble_sort(int * id, double * wage, int length);
double random_no();
int calculate_random_agent();
```

The example header file 'my_library_header.h' is included in one of the function file to compile it. Any global functions used by the model can also be defined as prototypes here.

4.8.2   Time Rules

Time rules allow restricting functions to act during particular iterations. An iteration refers to the smallest unit the model runs through in full cycle.

Most models can represent one day in the calender or one week depending on how modelers have designed them. The following depicts the use of time units in a model description file to declare various time periods used in the model,

Within the environment tag:

```
<timeUnits>
  <timeUnit>
  <name>daily</name>
  <unit>iteration</unit>
  <period>1</period>
  </timeUnit>
  <timeUnit>
  <name>population-regenerate</name>
  <unit>iteration</unit>
  <period>100000</period>
  </timeUnit>
</timeUnits>
```

Time rules are defined by a time period and a phase, defined as a time unit and an offset from start of a period. These can also be defined using a value from agent memory:

```
<condition>
  <time>
    <period>monthly</period>
    <phase>a.day_of_the_month_to_act</phase>
  </time>
</condition>
```

Modelers can define a function to perform at particular iterations (or specific days of the calender). Time rules are defined as conditions. The parser places these as rules in rules.c file. Example of a condition in agent function.

```
<function>
  <name>Actor_post_my_location</name>
  <currentState>00</currentState>
  <nextState>01</nextState>
  <outputs>
    <output><messageName>actor_location</messageName></output>
  </outputs>
  <condition>
    <time>
      <period>popBoard_start</period><phase>1</phase>
    </time>
  </condition>
</function>
```

Chapter 5

Agents in Social Science

5.1   Sugarscape Model 92
      5.1.1   Evolution from Bottom-Up 93
      5.1.2   Distribution of Wealth 94
      5.1.3   Location Is Important! 95
      5.1.4   Find Agents around Me 104
      5.1.5   Handle Multiple 'Eaten' Requests 105
      5.1.6   Change Starting Conditions 105
5.2   Modeling Social Networks 107
      5.2.1   Set Up a Recurring Function 112
      5.2.2   Assigning Conditions with Functions 113
      5.2.3   Using Dynamic Arrays and Data Structures 113
      5.2.4   Creating Local Dynamic Arrays 114
5.3   Modeling Pedestrians in Crowds 114
      5.3.1   Calculate Movement toward Other Agents 116
      5.3.2   Entering and Exiting Agents 118

Decentralized control is an important aspect of self-organizing systems. Behavior of insect colonies is studied to deduce how, despite working independently, a colony can work so efficiently. Termites and ants are examples of this. In ant colonies, Wilson and Hölldobler [202] argued that every ant follows a particular 'rule of thumb' while making decisions based on local stimuli. When all actions are put together, this emerges into complicated but precise execution plans for the colony. This behavior has evolved over millions of years, driven by natural selection.

Growing artificial agent societies is a useful technique to study how societies are created and thrive in changing real world conditions. Computer simulations can be used to study organisms interacting together in a safe environment, validated with experimental data.

Social scientists have used agent-based models in various political, ecological and economic scenarios. Here, agent-based models are ideal for understanding models involving individuals who interact and produce emergent phenomena. Writing agent-based models begins with assumptions on the interactions among agents. The agents are then simulated, producing and modifying variables depending on these interactions and time. Simulations are used as an addition to scientific analysis from deductions and inductions. Here simulations begin with a rigourous set of assumptions and generate data which are analyzed via induction.

Tesfatsion and Axelrod [194] discussed specific goals for agent-based researchers to pursue.

Empirical. How will particular large-scale phenomena perform such as emergence of social norms? These are analyzed to see if they pursue with global irregularities, to find why certain behaviors persist.

Normative. How are agent-based models used to discover good designs? Investigate key issues like efficiency and order.

Heuristic. Can complex behaviors be attained via simple interactions? It is difficult to predict behavior of large-scale systems with interactions on smaller scales.

Methodological Advancement. Study models and simulation data are produced.

Conway's Game of Life is one of the earliest examples using cellular automata grids, to display a group of cells interacting with each other [72]. Using only four simple rules, through simulation, the game could present new patterns depending on neighboring cells. For every cell, the four rules are

1. If current cell is alive and has less than two neighboring cells - die due to lack of social activity.

2. If current cell is alive and has more than three neighboring cells - die due to overcrowding.

3. If current cell has two or three live neighbors - survive to next time step.

4. If current cell is a dead cell but has three live neighbors - become alive.

.

Since this example, computer simulations have taken a long journey to more complicated computer programs using agent-based modeling and parallel computing. Figure 5.1 shows a snapshot of a game of life run. The black dots represent live cells. All cells follow the four simple rules, producing the various patterns.

Schelling [169] used a segregation model to show that, by adding a small preference factor, societies can emerge into segregated ones. The model, based on 2D grid landscape, had individuals represented by different colors. Every colored individual would check for the following rule:

If more than 33% of adjacent individuals or cells were of different color, the cell should randomly move to new position.

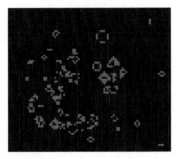

FIGURE 5.1: Snapshot of game of life during a simulation. Adapted from [141].

This model demonstrated that by just adding a small factor of 33% for neighbor preferences, societies would eventually separate out with time. However, if the preference factor were increased to 50%, the model would fail. Individuals would then have a 50-50 preference for neighbors, allowing societies to accept and consider staying.

One of the famous examples of a social agent-based model was written by Epstein and Axtell [57]. This model used an extremely simple setting for an agent society that played also simple rules, but showed complex practices. The experiment called 'Sugarscape model' was one of the earliest agent-based models that allowed researchers to investigate various aspects of society.

The model contained an artificial society of agents, who were allowed to move around on a 2D grid space to look for sugar. Figure 5.2 depicts screenshots of the model, before the simulation begins. The sugar was distributed in two piles on opposite corners of the grid with agents distributed randomly.

The traditional Sugarscape model allowed agents to see in four directions to search for sugar. Figure 5.3 shows the viewing distances of the agent. The agents are laid on a grid structure, using cellular automata. The agents can look north, south, east and west, ignoring the diagonally positioned cells. The agents use this rule to look for sugar and move to the sugar-laden square to eat it. With inclusion of location, the Sugarscape model showed how spatial distribution or landscape can influence the inhabitant agents and resources around them. The agents were seen crawling over the landscape, looking for sugar in Figure 5.4.

The sugar acts as a source of energy, distributed in two piles over the landscape. The agents also have a metabolism that uses up the sugar each time they move. The basic rules followed by the agents were

1. Agent looks north, south, east or west for sugar. The agent's vision was dependent on the modeler's code as to how many square lengths of vision are allowed.

2. If sugar is found, move to sugar location and eat.

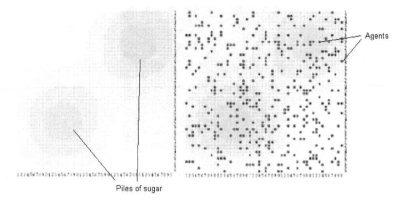

Piles of sugar

FIGURE 5.2: Initial distribution of sugar (left) and with agents (right). Adapted from [21].

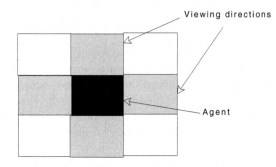

FIGURE 5.3: Agent perception. They can see north, south, east and west.

3. If sugar is not found, randomly move to another square.

Each time the agent moved, a small amount of its stored sugar was used by metabolism. Eventually, having used up all the sugar, the agents would die and disappear from the scenario.

Sugarscape allowed multiple research questions to be investigated by adding simple extensions to the model. This made it an ideal model replicating complex real societies. Researchers [102, 118] have extended their models with

Measuring wealth distribution. Sugars collected by agents were assessed to see what portion of the society was able to capture the most sugar.

Disease propagation. Diseased agents are introduced during the simulation to see it spread across the landscape.

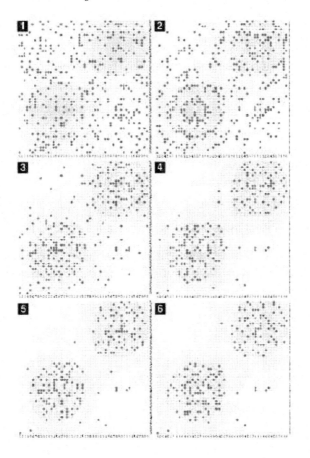

FIGURE 5.4: During the simulation, agents move to high sugar concentration areas. Adapted from [21].

Forming of social networks. Social networks are formed between neighboring agents or with whom they collided.

Migration among agents. Territorial areas are established, with agents moving to them, similar to how migration works in real world searching for jobs or resources.

Sexual reproduction. Agents were given certain genetic material. Each agent could scan neighbors, and choose a neighbor whose genetic material would be most similar to itself. Fertile agents were quickly seen to find each other and mate to produce new agents, sprouting across the landscape.

Inheritance among family members. With the concept of families, inheritance of wealth or sugar was programmed. This allowed various society classes to be formed.

Combat. Sometimes agents were allowed to combat each other for commodities.

Life and death. Agents were given life spans, modeling a living society.

Trading between sugar and spice. Wealth was denoted by the amount of sugar. Spice was an additional commodity, introduced by Epstein et al. [57]. The agents were told they need specific proportions of both sugar and spice to survive. During the simulation, if an agent had too much sugar and *bumped* into another agent, who had extra spice, the two agents would agree to *trade*. Through this model, supply and demand curves were generated, similar to Figure 6.6, useful for economic models.

Terrains. Some models had mountains, and agents concentrated in areas of high sugar that were easily accessible. Figure 5.4 shows agent movement across landscape in a simple Sugarscape model. The agents concentrated into two areas, where there were sugar piles in the beginning.

5.1 Sugarscape Model

The Sugarscape model proved to be an excellent tool to analyze economic models in an artificial society. Horres and Gore [92] discussed similarities between economics in real and artificial societies, using the spice trading facet. The authors presented the supply-demand curve with the equilibrium changing with different experimental settings. Gumerman et al. [79] used the model to produce results, that were later mapped to show how prehistoric American society settlements were organized in history. Klöck [109] presented a detailed report on Sugarscape model with territory formation. His work investigated behavior of trade, combat and wealth distribution when factors like migration and taxes were introduced.

Sugarscape's key advantage was that it could be tweaked with variables and methods to allow new behaviors. By introducing authorities or leaders, governments could be seen emerging. Peterson [148] supported the use of the model stating that "by providing insights into population growth, resource use, migration, economic development, conflict, and other global social processes, games played on the Sugarscape grid, may help shape policies needed to direct future course of society."

The Sugarscape model was also used in economics. Al et al. [3] used the model to study effects of taxing wealth and redistribution when they measured collected taxes over a population of 400 agents. Their results showed that using high tax rates was good for the population to survive, but poor agents struggled to survive [16, 3].

Buzing et al. [36] showed learning and communication influencing agents. Only certain agent populations were allowed to learn new strategies from other agents. The results showed that evolution only influenced those agents who listened. Increasing the communication among agents increased cooperation among the population, depicting that societies with no or little communication find it difficult to survive. These results are similar to those provided by Noble [142]. Hales [80] used the Sugarscape society with memetic algorithms to display propagation of cultural information among the population.

The Sugarscape was also heavily criticized in [168] saying that it was too restrictive to be used as an economic analysis model. This is because some results of the model showed lack of steady-state behavior in some scenarios. They claimed that as Sugarscape omitted existing economic theories, it could not be used for testing. Beinhocker [21] argued that Epstein and Axtell had not expected that Sugarscape would become a model for economics. Yet it was able to produce striking results free of unrealistic assumptions found in traditional economics. It is not based on an equilibrium system and neither does it go into it. It is a useful model which displays complex structures, evolving from bottom-up, using simple starting rules at low-level interactions.

Whether Sugarscape is a useful model for testing economic models is a debatable issue. However, it is a good starting point for modeling economic activities, where location of agents influences agent behavior. Most economic models do not have concepts of location to see if this affects agent and resource distribution.

5.1.1 Evolution from Bottom-Up

Sugarscape is also useful to see how societies develop and evolve. If agents were equipped with genetic material, this could be inherited by newborn child agents, allowing best genes to be carried onto new generations. Running such an experiment, "over a period of time, the characteristics of the population of agents converge towards certain traits, namely good vision and a low metabolism" [118].

Similar institutions, like banks show economic trading was an example of emerging evolution. This was observed, when Epstein and Axtell [57] introduced the following rules:

An agent could be a lender if it is too old to have children or has more savings than needed for reproductive purposes.

An agent could be a borrower if it has insufficient savings to produce children, but a sufficient supply of either sugar or spice.

An interest rate could be added on loans which can be collected.

If agents were credit worthy they can borrow sugar.

When Epstein and Axtell plotted the relationships between the lenders and borrowers, they were able to trace complex relationships between rich and poor agents. The relationships showed that rich agents were lending to poorer agents through middle agents who performed functions similar to banks behaving in real life (Figure 5.5).

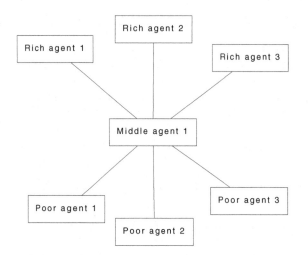

FIGURE 5.5: Relationships emerged between rich and poor agents. The middle agents behaved like banks.

As with the other emergent patterns in Sugarscape, the evolution of these credit networks was not in any way imposed from the top down on the model. Rather, these large-scale macro patterns grew from the bottom up, from the dynamic interplay of the local micro assumptions. [21]

5.1.2   Distribution of Wealth

Traditional economic theories follow Pareto laws, that markets always lead to perfect allocation of resources among the population. Beinhocker [21] used the Sugarscape model to display this pattern in sugar distribution across agents. Figure 5.6 shows cumulative distribution of sugars across agents. The figure supports the economic inequality among agents, displaying a right-hand tail stretching to have only a few rich agents. The *bump* formed in the middle in the beginning of the simulation slowly shrank as time progressed.

The model displayed no relation between the cause and effect, as to why some agents are poorer than others and what could be the reasons for this inequality.

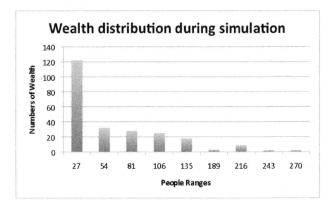

FIGURE 5.6: Wealth distribution among agents, with initial random sugar distribution. cf. [21].

Being an emergent property, these results supported the 'invisible-hand' phenomenon of economics. However, it is possible that initial distributions may affect how agents become wealthy. Agents lying closer to sugar areas become richer quicker than agents situated far away.

5.1.3 Location Is Important!

The Sugarscape model is a sociological model, where the position hold great influence on interactions. Each agent (citizen) has an area of vision and moving distance for capturing sugar. The decisions are made locally depending on what happens in this area of vision of agents.

Figure 5.7 shows an area of influence of one agent, where agent decisions are based on local stimuli. Agents make all decisions locally, on what they see in the circle of influence or bounded view. This makes the society highly decentralized, similar to ant or termite colonies in nature. The model can be analyzed, on a larger scale, for patterns emerging from the complete society. Table 5.1 summarizes the model parameters for distances considered in the FLAME Sugarscape model.

Table 5.2 describes how the model looks with agents, memory and functions. The citizen and sugar both need x and y coordinates located on a 2D landscape. The distance will allow closest citizens to eat sugar first. Once eaten, the sugar will 'die' or disappear from the scene. The messages allow

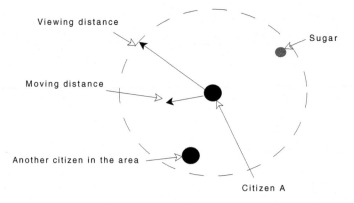

FIGURE 5.7: View of a citizen agent in FLAME Sugarscape.

TABLE 5.1: Global variables used in FLAME Sugarscape.

Viewing distance	200
Eating distance	5
Moving or run distance	5.5
Landscape	200 × 200

agents to see or find each other and post information to each other. Figure 5.8 displays the basic functions the agents perform in one iteration. The sugars have to post their location allowing citizens to find them. If the sugar is eaten by citizens, these disappear from the landscape. Note that the iteration timeline shown in Figure 5.8 does not consider citizen agents dying or trading sugar.

```
<!-- Model file for Sugarscape-->
<xmodel version="1">
<name>Sugarscape Model</name>
<version>1</version>
<author>Mariam Kiran</author>
<date>300809</date>

<environment>
  <functionFiles>
    <file>my_library_functions.c</file>
    <file>citizen.c</file>
    <file>sugar.c</file>
  </functionFiles>
</environment>
```

TABLE 5.2: FLAME Sugarscape model.

Identify Agents	Citizen, Sugar
Agent memory	Citizen: id, sugars, x, y; Sugar: id, x, y
Agent functions	Citizen functions: Citizen posts location, Citizen look for sugar, Citizen eats sugar, Citizen updates its sugar count; Sugar functions: Sugar posts location, Sugar checks it is eaten or not
Messages needed for communications	Citizen location: Contains citizen id, x, y; Sugar location: Contains sugar id, x, y; Eaten: Contains citizen id, sugar id

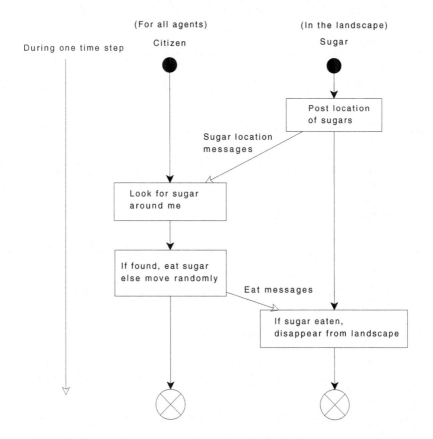

FIGURE 5.8: Timeline of the basic FLAME Sugarscape model.

```
<xagent>
<name>Citizen</name>
<description></description>
<memory>
   <variable><type>int</type><name>id</name></variable>
   <variable><type>int</type><name>sugars</name></variable>
   <variable><type>double</type><name>x</name></variable>
   <variable><type>double</type><name>y</name></variable>
   <variable><type>int</type><name>flag_sugar_found</name></variable>
</memory>

<functions>
   <function>
   <name>Citizen_post_location</name>
   <currentState>01</currentState>
   <nextState>02</nextState>
   <outputs>
      <output><messageName>citizen_location</messageName></output>
   </outputs>
   </function>

   <function>
   <name>Citizen_look_for_sugar</name>
   <currentState>03</currentState>
   <nextState>02a</nextState>
   <inputs>
      <input><messageName>sugar_location</messageName></input>
   </inputs>
   </function>

   <function>
   <name>Citizen_Eaten</name>
   <currentState>02a</currentState>
   <nextState>06</nextState>
   <condition>
      <not>
        <lhs><value>a.flag_sugar_found</value></lhs>
        <op>EQ</op>
        <rhs><value>0</value></rhs>
      </not>
   </condition>
<inputs>
  <input><messageName>eaten</messageName></input>
</inputs>
<outputs>
  <output><messageName>my_sugar</messageName></output>
</outputs>
</function>
   </functions>
```

```
</xagent>

<xagent>
  <name>Sugar</name>
  <description></description>
  <memory>
    <variable><type>int</type><name>id</name></variable>
    <variable><type>double</type><name>x</name></variable>
    <variable><type>double</type><name>y</name></variable>
  </memory>

  <functions>
  <function>
  <name>Sugar_post_location</name>
  <currentState>00</currentState>
  <nextState>01</nextState>
  <outputs>
    <output><messageName>sugar_location</messageName></output>
  </outputs>
  </function>

  <function>
  <name>Sugar_check_eaten</name>
  <currentState>01</currentState>
  <nextState>02</nextState>
  <inputs>
    <input><messageName>request_sugar</messageName></input>
  </inputs>
  <outputs>
    <output><messageName>eaten</messageName></output>
  </outputs>
  </function>

  </functions>
</xagent>

</agents>

<messages>
  <message>
    <name>citizen_location</name>
    <variables>
      <variable><type>int</type><name>citizen_id</name></variable>
      <variable><type>double</type><name>x</name></variable>
      <variable><type>double</type><name>y</name></variable>
    </variables>
  </message>
```

```xml
    <message>
      <name>sugar_location</name>
      <variables>
        <variable><type>int</type><name>sugar_id</name></variable>
        <variable><type>double</type><name>x</name></variable>
        <variable><type>double</type><name>y</name></variable>
      </variables>
    </message>

    <message>
      <name>eaten</name>
      <variables>
        <variable><type>int</type><name>citizen_id</name></variable>
        <variable><type>double</type><name>x</name></variable>
        <variable><type>double</type><name>y</name></variable>
      </variables>
    </message>

    <message>
      <name>my_sugar</name>
      <variables>
        <variable><type>int</type><name>citizen_id</name></variable>
        <variable><type>int</type><name>sugars</name></variable>
      </variables>
    </message>

</messages>
</xmodel>
```

```c
//Library file contains the global values used by the C
  functions
#define citizen_view_length 200.0
#define citizen_eat_length 5.0
#define citizen_run_length 5.5
#define CitizenBoardSize 10
#define LENGTH 4
#define THRESHOLD 10.0
#define LANDSCAPE 200
double handle_boundary(double position);
```

```c
//Library functions file contains common functions used by
  agents
#include "header.h"
#include "my_library_header.h"

double handle_boundary(double position)
```

```
{
  double new_pos=position;
  if(position >= LANDSCAPE)
  {
    new_pos =LANDSCAPE -(position-LANDSCAPE);
  }
  else if(position<0)
  {
    new_pos=new_pos*(-1);
  }
  else
  {
    printf("value is %f", new_pos);
  }
  return new_pos;
}


/**Citizen Agent functions file containing its actions**/
#include "header.h"
#include "my_library_header.h"
#include "Citizen_agent_header.h"

//Function to handle agent placement as a 2x2 Grid
double sec_handle_boundary(double position)
{
  double new_pos=position;
  if(position >= LANDSCAPE)
  {
    new_pos =LANDSCAPE -(position-LANDSCAPE);
  }
  else if(position<0)
  {
    new_pos=new_pos*(-1);
  }
  else
  {
    print("Possible check");
  }
  return new_pos;
}

int Citizen_post_location()
{
  add_citizen_location_message(ID, X, Y, SCENE_ID);
  return 0;
}
```

```
int Citizen_look_for_sugar()
{
  int closest_sugar_id=-1;
  double shortest_distance=9999.0;
  double current_distance_squared;
  double current_distance_squared_citizen;
  double closest_x,closest_y;
  int richest_citizen_id=-1;
  int citizen_sugars=0;
  int max_sugar=0;//change to 0

  FLAG_SUGAR_FOUND=0;

  sugar_location_message=get_first_sugar_location_message();
  while(sugar_location_message)
  {
    //Extracting information from message
    current_distance_squared=
        (sugar_location_message->x-X)*(sugar_location_message->x-X) +
        (sugar_location_message->y-Y)*(sugar_location_message->y-Y);

      if(current_distance_squared <=
                    (citizen_view_length* citizen_view_length))
      {
        if(current_distance_squared<shortest_distance)
        {
          shortest_distance=current_distance_squared;
          closest_sugar_id=sugar_location_message->sugar_id;
          closest_x=sugar_location_message->x;
          closest_y=sugar_location_message->y;
          FLAG_SUGAR_FOUND=closest_sugar_id;
          printf("Sugar is found");
        }
      }
    }
    sugar_location_message=get_next_sugar_location_message
                                    (sugar_location_message);
  }

  if(FLAG_SUGAR_FOUND!=0)
  {
    //move randomly in the space
    X=X+(citizen_run_length-((double)rand()/(double)(RAND_MAX)*
                        (citizen_run_length*2.0)));
    Y=Y+(citizen_run_length-((double)rand()/(double)(RAND_MAX)*
                        (citizen_run_length*2.0)));
  }
```

```
    X=sec_handle_boundary(X);
    Y=sec_handle_boundary(Y);
    return 0;
}

int Citizen_Eaten()
{
  eaten_message=get_first_eaten_message();
  while(eaten_message)
  {
    if(eaten_message->citizen_id==ID)
    {
      SUGARS++;
    }
    eaten_message=get_next_eaten_message(eaten_message);
  }
  return 0;
}

/**Sugar Agent functions file containing its actions**/
#include "header.h"
#include "my_library_header.h"
#include "Sugar_agent_header.h"

int Sugar_post_location()
{
  add_sugar_location_message(ID, X,Y);
  return 0;
}

int Sugar_check_eaten()
{
  int citizen_id=-1;
  request_sugar_message=get_first_request_sugar_message();
  while(request_sugar_message)
  {
    if(request_sugar_message->sugar_id==ID)
      {
        citizen_id=request_sugar_message->citizen_id;
      }
    request_sugar_message=get_next_request_sugar_message
                                 (request_sugar_message);
  }

  if(citizen_id!=-1)
  {
    add_eaten_message(citizen_id, X, Y);
```

```
    return 1; //returning 1 deletes the agent from the
      simulation
  }
  return 0;
}

<!-- Starting conditions with 1 citizen and 1 sugar-->
<states>
<itno>0</itno>
<xagent>
  <name>Citizen</name>
  <id>50</id>
  <sugars>0</sugars>
  <x>33.542894</x>
  <y>59.733879</y>
  <flag_sugar_found>0</flag_sugar_found>
</xagent>
<xagent>
  <name>Sugar</name>
  <id>1</id>
  <x>90.304270</x>
  <y>98.867763</y>
</xagent>
</states>
```

5.1.4 Find Agents around Me

Although sugar is an inactive entity, it had to be located and found by other agents. The sugar agents perform the following:

- Post their location messages, so they are read by citizen agents.

- Read 'eat' messages to determine if they are eaten and should disappear from landscape.

Agents can parse through messages to work out if there are agents close to it. The citizen agents read through a list of sugar location messages to collect their x and y coordinates. These allow a distance to be calculated and then measure the viewed distance to calculate if the citizen can see these sugars. If any sugar is found, the ID is recorded and then posted in the next loop to tell the sugar agent that it was found.

```
sugar_location_message=get_first_sugar_location_message();
while(sugar_location_message)
{
  //Extracting information from message
  current_distance_squared=
```

```
           (sugar_location_message->x-X)*(sugar_location_message->x-X) +
           (sugar_location_message->y-Y)*(sugar_location_message->y-Y);

  if(current_distance_squared <=
                      (citizen_view_length* citizen_view_length))
  {
    if(current_distance_squared<shortest_distance)
    {
        FLAG_SUGAR_FOUND=closest_sugar_id; //Agent memory saves this id
        //"Sugar is found"
    }
  }
}
  sugar_location_message=get_next_sugar_location_message
                                  (sugar_location_message);

}
```

The next function can then use its memory values to determine if it needs to interact with sugar agents in the iteration.

```
<condition>
  <lhs><value>a.flag_sugar_found</value></lhs>
  <op>EQ</op>
  <rhs><value>0</value></rhs>
</condition>
```

5.1.5 Handle Multiple 'Eaten' Requests

By recording the ID of the sugar agent closest to the citizen agent, the citizen can then choose which sugar to interact with. When the sugar agents then read these 'request' messages, they can choose one of the agents and be 'eaten' by one.

5.1.6 Change Starting Conditions

The experiment can be repeated with different starting conditions, by changing values defined in 0.xml. Figure 5.9 displays the three initial settings of agents, varying x and y positions of agents.

Figure 5.10 represents the sugar distribution captured by citizens in the simulation. The sugar distribution is shown in Figure 5.11 as frequency logs.

The results display a fairly egalitarian society where the wealth distribution is a smooth bell-shaped curve. Starting with a small number of rich and poor agents, a broad middle class emerges, with a small distance between the rich and the poor in the beginning. However with time, fewer citizen agents emerge as the super rich and the middle class shrinks, increasing poor agents. Initial conditions on locations also influenced wealth distribution, with citizen

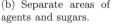

(a) Random mixed distribution of agents and sugar.

(b) Separate areas of agents and sugars.

(c) Overlapping areas of sugar and agents.

FIGURE 5.9: Three different initial settings for simple Sugarscape experiment. The citizen agents are represented by *red* dots and *green* dots represent sugar agents in the scenario.

agents being quick to grab sugars close to them, displaying a positive skewness with virtually no poor agents.

> Skewness is the measure of asymmetry of data distribution. Kurtosis is the measure of the bulge or peakedness of data distribution.

Table 5.3 summarizes the skewness and kurtosis of sugars collected at time $t = 500$ during the simulation. The higher the kurtosis, the greater the distribution between the rich and poor. The experiment with separate areas of location for both citizens and sugars displayed the highest difference between rich and poor. This was followed by the random distribution and lastly the overlapping area results.

Figure 5.11 displays a cumulative distribution of the sugars gathered. The maximum amount of sugar held by anyone of the 50 citizens was 120.

$$\frac{20}{100} \times 120 = 24 \tag{5.1}$$

Figure 5.11(b) depicts a pattern similar to Pareto layout, with cumulative distribution displaying an 80% of wealth held by 20% of the population. The plots for the three experiments showed that most sugar was captured between the 20-30 agent distribution. However, the experiments for overlapping and separate areas took longer to achieve this. This shows that the Pareto law holds for random distributions of agent wealth, but when initial conditions are varied their distributions changed.

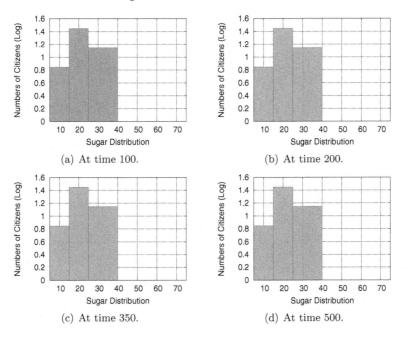

(a) At time 100. (b) At time 200.

(c) At time 350. (d) At time 500.

FIGURE 5.10: Sugar collected for random initial agent distribution.

5.2 Modeling Social Networks

Social networks emerge through local interactions among individuals. These help form local networks or groups. Agent-based models can be used to study these social networks by programming and visualizing how bonds are formed and broken. Emergence of these networks can thus be simulated, to study why people make and break contacts.

Snijders et al. [190] used actor-oriented models and rules to show networks forming over time. Different bonds were influenced by actor decision making, measuring factors such as number of outdegree, instrumental and social ties. Prell [151] used FLAME to model social capital and network formation, forming ties based on gains and job positions. Shown in Figure 5.12, a basic network was seen to evolve over time. The agent-based model contained a collection of heterogeneous actors, with different memory variables, each making decisions on a set of rules to form a tie or not. These rules took into account the cost of tie formation and eventual gain of forming them.

Table 5.4 describes the model specifics with agents and functions involved. The model was analyzed to understand the degrees of centrality, star structures and average wealth gained by agents.

The model involved only one kind of agent, *Actor*, with different memory

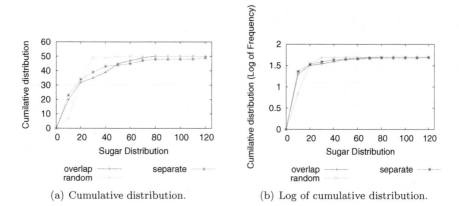

(a) Cumulative distribution.　　　(b) Log of cumulative distribution.

FIGURE 5.11: Distribution of captured sugar.

TABLE 5.3: Results of skewness and kurtosis measures in three experiments.

Initial Distri-bution	Random	Separate Areas	Overlapping Areas
Skewness	1.586	2.418	1.530
Kurtosis	2.047	6.043	1.692

values but same functions. The memory variables included factors like number of projects, number of ties, who is in my circle and agent interests. These variables were used in communication with other actors to find similarities (to make ties) or compete in projects (to break ties). The agent functions, thus, involved reading the actors close by, and making decisions on whether to make or break a tie with them. Over time, various relationships were formed which were either direct, reciprocative or transitive bonds.

The model involved simulations with over 1000 actors, and took about 30,000 iterations to stabilize. The results showed levels of reciprocity, similarity and transitivity affected the actors leading to higher clustering in the networks. However, the transitive relationships produced higher effects on the social well-being of the actors. And open two-star network structures reduced the amount of clustering, affecting the direct relationship with outdegree tie formations (Figure 5.13).

Modeling social and economic worlds usually leads to show presence of an equilibrium, when the society is at a maximum benefit, utilizing all resources and performing efficiently. Social capital theory can be useful to study how resources in social networks can be used to study network formation, through social and instrumental ties [119, 136]. Simulation outputs can be analyzed for distribution of social resources, money and other actor attractions when

(a) Network at Time (t=1).

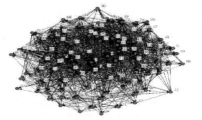

(b) Network at Time (t=500).

(c) Network at Time (t=1000).

FIGURE 5.12: Evolution of networks in a simulation. Adapted from [151].

forming network structure. These can be the base for more complex studies of how resource is eventually distributed and social inequality emerges. This also involves cost-benefit calculations bringing back the notion of rational actors, making correct decisions for maximum gain.

Following code snippets from the model, show how the actors making ties based on the following information:

- For every actor calculates how many actors are around me within viewing distance.

- For every actor I have a tie with, calculate if I have to compete and break the bond.

- For every actor I do not have a tie with, calculate if I should form one.

```
<!- In model file, define a data structure to help document ties->
<xmodel>
.....
<functionFiles>
  <file>my_library_functions.c</file>
  <file>actor.c</file>
</functionFiles>

<timeUnits>
  <timeUnit>
  <name>daily</name>
  <unit>iteration</unit>
```

TABLE 5.4: Social network model specifications.

Agents	Actor
Actor memory	Id, number of ties, number of projects, payoff, pathlengths, degree of centrality, clustering and similarity, well-being factor, cost
Actor functions	• Post my ties and project details • Calculate total ties I have • On random, add or drop a tie • Calculate total connections
Messages for communication	Actor ties, actor locations, actor wealth

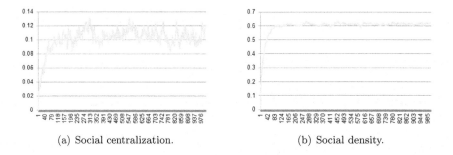

(a) Social centralization.   (b) Social density.

FIGURE 5.13: Evolved centralization and density in networks. Adapted from [151].

```
    <period>1</period>
  </timeUnit>

  <timeUnit>
    <name>popBoard_start</name>
    <unit>iteration</unit>
    <period>100000</period>
    </timeUnit>
</timeUnits>

<environment>
  <dataType>
  <name>bond</name>
  <variables>
    <variable><type>int</type><name>from</name></variable>
    <variable><type>int</type><name>to</name></variable>
```

```
    </variables>
    </dataType>
</environment>

<!-- The bond data type can be part of the agent memory-->

<xagent>
  <name>Actor</name>
  <memory>
  <variable><type>int</type><name>id</name></variable>
  <variable><type>int</type><name>perform</name></variable>
  <variable><type>int</type><name>project_flag</name></variable>
  <variable><type>bond_array</type><name>mybonds</name></variable>
  <variable><type>int</type><name>knowledgecategory[10]</name></variable>
  <variable><type>int</type><name>money</name></variable>
  <variable><type>int</type><name>number_of_transitive_ties</name></variable>
  <variable><type>int</type><name>x</name></variable>
  <variable><type>int</type><name>y</name></variable>
  </memory>

  <functions>
    <function>
    <name>Actor_post_my_location</name>
    <currentState>00</currentState>
    <nextState>01</nextState>
    <outputs>
      <output><messageName>actor_location</messageName></output>
    </outputs>
    <condition>
      <time><period>popBoard_start</period><phase>1</phase></time>
    </condition>
    </function>

    <function>
    <name>Actor_update_my_bonds</name><description></description>
    <currentState>01</currentState>
    <nextState>01a</nextState>
    </function>
    ...
</xagent>
...
</xmodel>

\*Extract from actor.c file describing the main functions**/

int Actor_post_my_location()
{
  add_actor_location_message(ID,X,Y);
  return 0;
}

int Actor_update_my_bonds()
{
  bond_array bond_list;
  init_bond_array(&bond_list);
  int i;
  int actor_found =0;
```

```
double rnum;
rnum=(double)rand()/(double)RAND_MAX;

if (rnum<0.5) //break a tie
{
  for(i=0,i<MYBONDS.size;i++)//make a local copy of bonds
  {
    add_bond(&bond_list,MYBONDS.array[i].from, MYBONDS.array[i].to);
  }
  random_position=(double)rand()/(double)RAND_MAX*MYBONDS.size;
  //remove bond from a random actor
  remove_bond(&MYBONDS,random_position);

}
else // form a tie with an actor Im not connected to
{
    actor_location_message=get_first_actor_location_message();
while(actor_location_message)
{
    for(i=0;i<bond_list.size;i++)
    {
      //check if I already have a bond with another actor
      if(bond_list.array[i].from==ID) AND
                      (bond_list.array[i].to==actor_message->id)
      { //ignore
        actor_found =0;
      }
      else
      {
        actor_found =actor_message->id;
      }
      }//end of for loop
      actor_location_message=get_next_actor_location_message
                                  (actor_location_message);
  }
  if(actor_found!=0)
  {
    add_bond(&MYBONDS, ID, actor_found);
  }
  return 0;
}
```

5.2.1   Set Up a Recurring Function

Time units can allow functions to run only at certain iteration steps. The complexity of calculating this is handled by FLAME, but needs to be defined by modelers in the model XML file. In the example, using popBoard unit and setting its frequency to 10,000 means that this function will run every 10,000 steps in the simulation. The function Actor_post_location() should then add this time condition. This function will thus run once in iteration 1, and then only run every 10,000th iteration.

5.2.2 Assigning Conditions with Functions

```
<!-- in the model XML file-->
<function>
  <name>Actor_idle</name>
  <currentState>04</currentState>
  <nextState>04a</nextState>
  <condition>
    <lhs><value>a.function_perform</value></lhs>
    <op>EQ</op>
    <rhs><value>0</value></rhs>
  </condition>
</function>

<function>
  <name>Actor_project_calculation</name>
  <currentState>04</currentState>
  <nextState>04a</nextState>
  <condition>
    <lhs><value>a.function_perform</value></lhs>
    <op>NEQ</op>
    <rhs><value>0</value></rhs>
  </condition>
  <inputs>
    <input><messageName>actor_knowledge_expert</messageName>
     </input>
  </inputs>
</function>
```

Further complexity can be added to the XML file by associating conditions with functions. The above two functions branch from the same point, but run depending on the value of the memory variable 'function_perform'. If the value is equal to zero, the function 'idle' is run, else the 'project_calculation' runs. This mechanism can prevent modelers to add conditions in the C functions by handling them here.

5.2.3 Using Dynamic Arrays and Data Structures

The example also shows data structures defined and used in agent memory. This allows more complex memory variables to be created, such as to recording plans or records. The MYBONDS variable allows the actor to record its ties (from itself, to other actors), in one memory variable. This is defined as a dynamic array, allowing the number of ties to grow, rather than be a defined maximum (as when using static arrays).

5.2.4   Creating Local Dynamic Arrays

```
//Extract from a function showing a local int dynamic array
int_array actor_view;

//Calculate the actors in my area that I am not connected to
actor_location_message=get_first_actor_location_message();
while(actor_location_message)
{
  if(actor_location_message->actor_id!=ID)
  {
    for(i=0;i<MYBONDS.size;i++)
    {
      found=0;
      if(MYBONDS.array[i].to==actor_location_message->actor_id)
      {
        //if I am already connected
        found=1;
      }
    }
    if(found==0)
    {
      //if not found then add to my local list
      add_int(&actors_view, actor_location_message->actor_id);
    }
  }
  actor_location_message=get_next_actor_location_message
                         (actor_location_message);
}
free_int_array(&actors_view);
```

The local integer array is created by 'int_array actor_view'. These arrays are useful to record information locally, for immediate calculation where total number of variables is not known. It is important to free all arrays used at the end of the function, otherwise the memory allocated to them does not automatically free itself.

5.3   Modeling Pedestrians in Crowds

Crowd modeling can help study crowd behaviors in situations to help construct or plan safe building and pathways, or to include collision avoidance in panic situations or motion planning for exit planning in crowded venues. Thalmann [124] described the difference in individual person behavior as opposed to when they are part of a crowd. Sociological and behavioral simulation of people in closed environments allows studying relationships between different people from a social perspective. This can often display the existence of hierarchy inside a group, such as leadership and member relations among the individuals. Most crowd models have investigated behavior of one pedestrian within a crowd. However, group pattern formation and other types of social relationships can be extensively studied in crowds using agent-based models. Early examples can be seen by Reynolds [160], where he presented distributed behavioral model to produce flocking behavior.

Person agents in crowds can be goal-directed, reactive or opportunistic. Modelers have to program this preference into agents. Modeling crowd behavior requires a large amount of data analysis with different densities, numbers and heterogeneous behaviors. Most crowd behavior analysis is done using video tracking software and hindered by additional entities in scenes such as loose clothing, carrying umbrellas, bags or packages.

In FLAME, a crowd model was programmed as follows:

1. Initialize agents in scene (such as at entrance of a corridor).

   - Initialize 50 agents.
   - Randomly add people in a group to generate various group sizes.
   - Randomly generate families and assign ages between 4 to 100.
   - In groups, assign a leader.
   - Assign a destination exit for each agent group.

2. Post agent location for other agents to read.

3. Choose one of the following behaviors at random:

   - If agent is in group, collision avoidance to mediate movement.
   - If the agent is out of the group's circle, bring agent back towards group.
   - If in a group, the agent follows the leader.
   - The agent goes to nearest shop.

4. Walk all agents towards the exit.

The pedestrians are simulated as individual person agents walking in the crowd.

Person Agent. Contains id, x, y, speed, gender, exit_no, is a leader or not, which group do I belong to and other variables such as weights associated to prevent collisions and walk towards goals.

Generator Agent. Contains the total number of persons generated. This agent is necessary to create agents at corridor entry points in the simulation.

Corridor Agent. Contains locations of the corridor to allow the people to walk along the pathways. Sends information to all persons about wall locations to steer them in the right direction towards exits.

5.3.1   Calculate Movement toward Other Agents

```
int follow_leader()//calculate force to move towards leader
{
  int leader_group_member=-1;
  double shortest_distance=99999.0;
  double current_distance_squared=0.0;
  double leader_x, leader_y;

  double leader_velx,leader_vely;
  double leader_personx;
  double leader_persony;

  double myvel,myxy;
  double leader_personvel;
  double diffxy,diffvel;
  double posDiff;
  double springForce,dotRoot;
  double restLength=0.7;
  double strength=0.45;
  double damping=0.5;
  double dampingForce;

  //check all person location messages to find group members
  send_person_location_message = get_first_send_person_location_message();
  while(send_person_location_message)
  {
    if(send_person_location_message->group_id==GROUP_ID)
    {
      if((send_person_location_message->is_leader==1)
              &&(send_person_location_message->id!=ID))
      {
        current_distance_squared=
        (send_person_location_message->x-X)*(send_person_location_message->x-X)+
        (send_person_location_message->y-Y)*(send_person_location_message->y-Y);
        if(current_distance_squared < shortest_distance)
        {
          shortest_distance = current_distance_squared;
          leader_group_member = send_person_location_message->id;
          leader_x =send_person_location_message->x;
          leader_y = send_person_location_message->y;
          leader_velx = send_person_location_message->velx;
          leader_vely = send_person_location_message->vely;
        }
      }
    }
    send_person_location_message =
            get_next_send_person_location_message(send_person_location_message);
  }

  if(leader_group_member != -1)
```

```
{
    diffx=myx-leader_personx;
    diffy=myy-leader_persony;

    diffvelx=myvelx-leader_velx;
    diffvely=myvely-leader_vely;

    //using vector equations calculate the dot root function of the difference
    dotRoot=sqrt(dot(diffxy,diffxy));

    //posDiff is a data structure of two variables x and y
    if(dotRoot!=0.0)
    {
        posDiff=normalize(diffxy); //normalise the difference
        springForce=-(dotRoot-restLength)*strength;
        dampingForce=-damping*dot(posDiff,diffvel);//add damping to slow down

        posDiff*=(springForce+dampingForce);
        posDiff*=0.1;

        X+=posDiff.x;
        Y+=posDiff.y;
    }
}
return 0;
}
```

The new x and y positions are calculated using principles from vector transformations and force movements. The dot product between two vectors $A \cdot B$ calculates the new cartesian positions in the Euclidean vector space.

$$A \cdot B = \sum A_i B_i, \, for \, all \, i \, to \, n \qquad (5.2)$$

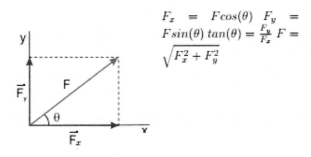

$$F_x = F\cos(\theta) \quad F_y = F\sin(\theta) \quad \tan(\theta) = \frac{F_y}{F_x} \quad F = \sqrt{F_x^2 + F_y^2}$$

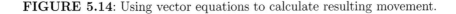

FIGURE 5.14: Using vector equations to calculate resulting movement.

Figure 5.14 summarizes calculations involved in calculating next positions x and y for agents. The agent will calculate, based on its own velocity and its leader's velocity vector, to get a resulting vector. The goal weights and damping force can allow the agent to move towards the leader, in the direction, giving the impression of a simple walk rather than a jump towards the leader. Similar mathematical equations can be updated to add more complexity in person movements, to make them appear more real.

5.3.2   Entering and Exiting Agents

```
//from the Generate Agent functions
int generate_people()
{
  int gnumber=rand()%MAXIMUM_GENERATED_PERSONS;
  int i=0;
  int leader=0;
  NUMBER_GROUP_GENERATE=NUMBER_GROUP_GENERATE+1;

  for(i=0; i<gnumber;i++)
  {
    if(i==0)
    {
      leader=1;
    }
    else
    {
      leader=0;
    }

    NUMBER_PEOPLE_GENERATE=NUMBER_PEOPLE_GENERATE+1;
    add_person_agent(NUMBER_PEOPLE_GENERATE,  X, Y, X,Y, 5.0, 1, 0,0,1,
                     NUMBER_GROUP_GENERATE,leader);
  }
  return 0;
}
```

The Generate agent can add new Person agents into the scene, using the function 'add_person_agent()'. The agent will first have to calculate the variables for the person memory and declare them as function parameters when creating agents.

For exiting agents, the person agent needs to have a function which returns a '1' rather than a '0', to allow it to be removed from the simulation. This can accompany an 'if' condition to check, if the exit point is reached, such as

```
if(current_location==exit_position)
  return 1;
else
  return 0;
```

Figure 5.15 shows two snapshots from the simulation, where the person agents are superimposed into a screen to walk around a collection of buildings.

(a) Single agents.   (b) Agents in groups.

FIGURE 5.15: Agents walking in the scene.

Figure 5.15(a) shows single person not associated in groups as compared to Figure 5.15(b) showing groups represented by different colors.

Various tools now exist that allow crowds to be modeled in different situations. These include VICrowd, Legion, each allowing multiple rule interactions and control to be introduced in the model. Designing such models requires considerable input from modelers, social scientists and crowd researchers to build believable crowds. The more complexity of lower and higher levels on interaction, the more believable crowds are.

Chapter 6

Agents in Economic Markets and Games

6.1   Perfect Rationality versus Bounded Rationality 125
6.2   Modeling Multiple Shopper Behaviors 126
6.3   Learning Firms in a Cournot Model 129
       6.3.1   Genetic Programming with Agents 143
       6.3.2   Filtering Messages in Advance 150
       6.3.3   Comparing Two Data Structures 151
6.4   A Virtual Mall Model: Labor and Goods Market Combined 152
6.5   Programming Games 159
       6.5.1   Nash Equilibrium 160
       6.5.2   Evolutionary Game Theory 161
       6.5.3   Evolutionary Stable State 162
       6.5.4   Game Theory versus Evolutionary Game Theory 162
       6.5.5   Continuous Strategies 163
       6.5.6   Red Queen and Equilibrium 163
6.6   Learning in an Iterated Prisoner's Dilemma Game 164
6.7   Multi-Agent Systems and Games 173

Economics, similar to social science, also uses mathematical concepts to help analyze and predict behavior of its systems. Traditionally, economics used differential equations, with various assumptions, including arguments of rational individual behavior and rational decision making. Game theory and economics, however, work hand-in-hand to help study people behavior and introduce concepts of payoff and utility when studying economics systems in research. Figure 6.1 shows economic models often viewed as black boxes, using inputs to then observe and collect its outputs. Researchers often work backwards to explain the output behavior using mathematical notations.

Economic agent-based modeling is a separate research area used to explain the inner workings of economics. Tesfatsion [193] defines agent-based computational economics (ACE) as "the computational study of economic processes modeled as dynamic systems of interacting agents. Here 'agent' refers broadly to a bundle of data and behavioral methods representing an entity constituting part of a computationally constructed world."

In economics the definition of an agent can vary from representing a group of agents, such as a firm composed of many individuals, or an individual itself

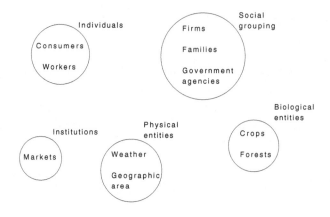

FIGURE 6.1: A black box represents an economic model where only inputs and outputs are known and little is known about what goes on inside.

FIGURE 6.2: Groups in economic systems.

like a customer or a worker (Figure 6.2). Agent-based modeling is solely based on an emergent pattern of interactions among different agents involved. Similarly, economies are also based on behavior of each member and the interacting patterns of these members.

The black boxes in economic models can thus be replaced with boxes full of agents (Figure 6.3). The agents can represent themselves or be used to represent a group of agents, where the interactions on lower levels affect their performance in the upper layers. Agent-based models produce output variables which are a result of interactions between agents within different scenarios linked up together. The earliest use of agent-based models can be found in the works of Axelrod [14], where he studied the evolution of cooperation among agents using the iterated prisoner's dilemma. Table 6.1 summarizes the main differences between traditional and complexity economics.

Each economic model is different, based on different perspectives and assumptions of modelers or economists,

Variables. Each model is made up of variables and equations. These models help understand the economy. If any one of the variables change the model changes. Examples of changing variables in models are estimating

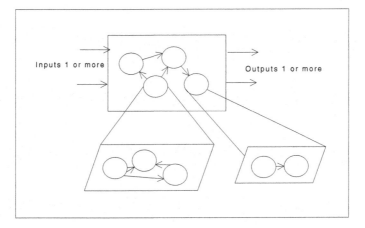

FIGURE 6.3: Replacing the black box with agents.

with or without expenditures, floating or fixed exchange rate or even increases in interest rate.

Limits. Every economic model has limits to what it is modeling.

Different kinds of consumers. Some consumers might be lazy while others spend more than required on products. There is heterogeneous mixture of characteristics in the real world.

Testing. Designing a test suite for testing different assumptions. This involves period testing where variables were the same for periods 1 and 2 but changed in period 3. Tesfatsion [193] argues that most models get rejected due to this.

Rules. Rules are determined out of some formulation of the past. These rules should be continually updated using learning methods. These learning methods will be conditional to the agents.

Behavioral uncertainty and learning in agents. Economic analysis how agents make choices in an evolving world. Holland et al. [90] argued why most economists turn to game theory to model strategic learning in games as economic games.

The SanteFe Institute presents their view on economic models [21]:

- Economic models are dispersed with parallel interaction among heterogeneous agents. Heterogeneity implies that each individual is different from the other in terms of memory and characteristics.

- There is no global entity which controls their functions.

TABLE 6.1: Five big ideas that distinguish complexity economics [21].

	Complexity Economics	**Traditional Economics**
Dynamics	Open, dynamic, nonlinear systems, far from equilibrium	Closed, static, linear systems in equilibrium
Agents	Modeled individually; use inductive rules of thumb to make decisions have incomplete information; are subject to errors and biases; learn and adapt over time	Modeled collectively; use complex deductive calculation to make decisions; have complete information; make no errors and have no biases; have no need for learning or adaptation (are already perfect)
Networks	Explicitly model interactions between individual agents; networks of relationships change over time	Assume agents only interact indirectly through market mechanisms (e.g. auctions)
Emergence	No distinction between micro and macroeconomics; macro patterns are emergent result of micro level behaviors and interactions	Micro and macro economics remain separate disciplines
Evolution	The evolutionary process of differentiation, selection and amplification provides the system with novelty and is responsible for its growth in order and complexity	No mechanism for endogenously creating novelty or growth in order and complexity

- Sometimes there is a hierarchy among the agents.

- There is learning in the agents as time progresses.

- Due to certain factors sometimes new market niches are seen developing.

- Importantly, economic models try to work away from the optimum or equilibrium because they are constantly trying to do better and never know whether they have reached an optimum point.

Dopfer argued that "economics has always been in a crisis since it broke away from social philosophy in the late eighteenth century" [52]. Since Aristotle's time, economic theories have changed a number of times, when Aristotle originally discussed the nature of household and market exchanges which concentrates mostly in political economics branch. Adam Smith's publication of *An Inquiry into the Nature and Causes of the Wealth of Nations* contributed to the discussion of free market which was much celebrated by economists thereafter. Smith argued that people's personal relationships contribute to the way markets behave [185]. The theory of 'invisible hand' encourages the laissez-faire policy adopted by most governments that allows events to take their own toll and have less interference with behavior of markets as they shape themselves.

Similar theories were adopted by neoclassical economics which gave birth to rational consumers and buyers, assuming every individual is making the right choice to maximize their own utility or profit. Conventional models of markets used assumptions of this 'rational choice' and 'efficient market hypothesis', but were limited to explain real market performance in situations of trading and volatility as observed in the real world.

6.1 Perfect Rationality versus Bounded Rationality

Friedman [70] presented ideas around how exaggerated assumptions will not matter in economics when the economic models being written were making correct predictions. Even if individuals were assumed to be perfectly rational, it would not make any difference on the results if they were making irrational decisions. Comparatively, Simon presented a counter argument on bounded rationality.

> "Economics illustrates well how outer and inner environments interact and, in particular, how an intelligent system's adjustment to its outer environment (its *substantive rationality*) is limited by its ability, through knowledge and computation to discover appropriate adaptive behavior (its *procedural rationality*)." [179]

Every individual is selfish and the information each individual has is different. The decisions are made, based on *what* the individual knows, giving rise to *bounded rationality*, where there is rationality depending on the bounds of the individual's information space.

6.2   Modeling Multiple Shopper Behaviors

A simple shop/customer model can be used to depict five different kinds of customers with different abilities, to study their behavior on the market. The agent description is as follows:

Agent - Shops. The shop agent sets new selling price of goods and posts to the message board. The customers would then buy these goods and, depending on the profit made by the shop, it would either raise or reduce the selling price in the next iteration. The shop keeps track of the income it receives after selling goods.

Agent - Customers. There are five different kinds of customers:

- Random shoppers (Type A): This type of customer will buy from any shop on random, without any previous knowledge.

- Customers who go to a favorite shop (Type B): These customers depend on old values while buying from one shop. If they are satisfied from the shop, they will go to the same shops to buy, else they choose another shop randomly.

- Customers who go to favorite shops of others (Type C): These customers depend on messages being posted by other customers. They then choose those shops and buy from those.

- Shoplifters (Type D): These customers are shoplifters who choose from any one shop and shoplift products.

- Customers who only buy from cheap shops (Type E): These customers will sort the shop list to find the cheapest shops and then buy from it.

The algorithm of the model during the simulation is:

1. Shop checks profits, sets good's selling prices and posts message 'open for business'. The customers calculate their wages and add these to their savings.

2. Customers then spend their savings, buying goods based on the shop price, stock message, and send updated stock message to shops.

3. Shops collect profits and add income on sold goods.

Based on the model description above, functions are as follows:

Function - Shop_1:   • Check the amount of stock sold in last iteration.

- If the stock sold is more that 5, increase the selling price by a random amount; else reduce it, check price does not go below zero.

- Post message of good's price.
- Post message of stock left in shop. Stock is by default started from 100.

Function - Customer types (A-E) 1:   • Calculate and set random wage.

Function - CustA_2:   • Calculate a random shop.

- Check if it has savings, buy goods from the shop.
- Post out new stock of the shop.
- Set new satisfied value if more than 0.5, post out message of satisfaction of the shop bought from.

Function - CustB_2:   • Check the satisfying value (past) of this customer. If more that 0.5, then the customer will buy from the past shop. Else calculate a shop on random and buy from that shop.

- Checks if savings exist, buy goods from the shop.
- Post out new stock of the shop.
- Set new satisfied value, if more than 0.5, post out message of satisfaction of shop bought from.

Function - CustC_2:   • Check the posted satisfied messages, get the first message and choose to go to that shop.

- Check if savings, buy goods from the shop.
- Post out new stock of the shop.
- Set new satisfied value if more than 0.5, post out message of satisfaction of this shop bought from.

Function - CustD_2:   • Choose a shop randomly.

- Check if shop has stock. If it does, then shoplift.
- Post out new stock of the shop.

Function - CustE_2:   • Choose 5 shops at random.

- Sort the shop list in order of cheapest.
- Check if savings, buy goods from the shop on top (cheapest shop).
- Post out new stock of the shop.
- Set new satisfied value if more than 0.5, post out message of satisfaction of this shop bought from.

Function - Shop_3:   • Find the latest stock message of shop agent and calculate how much stock was sold in this iteration.

- According to the stock sold and price, calculate income shop collected.

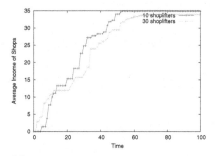

(a) Savings of the customers with all customer types to be of a population 10 each.

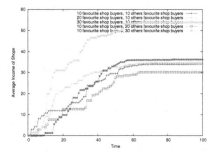

(b) Comparing the average income of the shops with 10 and 30 shoplifters.

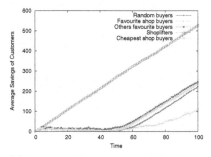

(c) Comparing the average income of the shops with 10 and 30 random buyers.

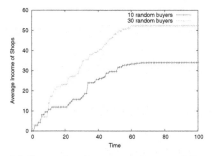

(d) Comparing the average income of the shops by varying the number of customers who go to their own favorite shops and those who go to shops recommended by others.

FIGURE 6.4: Different shoppers in the same simulation.

Figure 6.4(a) shows the average savings of all customer types. The shoplifters, since not spending save vast amounts compared to the others. The customers who keep buying from their favorite shop, despite of price increases, save the least in all five types. The other three customer types seems to save equally, even if they buy from cheap shops, randomly or buy from other recommended shops. Figure 6.4(b) compares the average income of shops with number of shoplifters in the system. The shops earn and save money with lesser shoplifters, than with more shoplifters. Figure 6.4(c) compares the average income of shops with number of random buyers. When there are more customers who buy at random, the average income of the shops is seen to increase. Figure 6.4(d) compares the average income of shops, if the customers who buy from their favorite or recommended by others is varied. If customers keep buying from their favorite shops, the shops are seen to earn more as the prices increase. But customers who buy on recommendations, seem to be loosing the most. This shows shops profit more by being recommended by other customers.

6.3 Learning Firms in a Cournot Model

The Cournot model is an economic mathematical model, implies a centralized model and works similar to the demand and supply curve. All firms produce a quantity of one type of good, where price is a central variable which connects all firms. Companies are constantly competing against each other over product sales and product demands in a market scenario trying to keep their profits high (Figure 6.5). Thus the price of a product affects all transactions and can be viewed as a central mediator in market scenarios. However, this only applies for homogeneous goods, ignoring products with wide variety and quality that influence market sales.

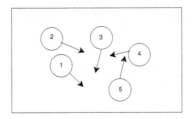

FIGURE 6.5: Five firms producing a particular output of the same product in the scenario.

Adopting this theory, various mathematical models have been developed in traditional economics to portray this concept of competition.

Cournot competition model. Firms make decisions about quantity of a product at each time step. These decisions are made concurrently and independent of other firms. Depending on the production and the product demand in the system, the price of the product changes over time.

Stackelberg competition model. Each firm takes turns to act as a leader and makes a decision on its production level. It is similar to playing an extensive form game with a decision tree in game theory. The strategies can then be represented as a decision flow showing firms making decisions one after the other [83].

Bertrand competition model. Similar to the Cournot model in assumptions and design, Bertrand firms decide how much they want to produce in the beginning and do not change their production throughout the simulation. Only the price is changed to adjust the profits collected by the firms [22].

The models are based on mathematical equations and carry large numbers of assumptions to work in practise.

- All production is sold even if it is given away for negative prices.

- These models of competition are all theoretical models involving mathematical calculations to explain the firm behaviors.

- These models come close to explaining the emergence of monopolies when one firm dominates the market through changes in supply and demand in real market behavior.

- All models assume an equilibrium which all firms will strive to achieve.

The Cournot [44] model is a simple economic model involving firms competing against each other for quantities they produce, to achieve high profit. The firms produce one homogeneous product and based on the demand in the system the price of the product changes. The characteristics of a supply and demand curve are shown in Figure 6.6. When the *supply* of product increases, it reduces the *demand* as there is an increase in abundance in the system.

The supply and demand curves have an inverse relationship with each other. The point '*E*' at which the two curves intersect is the equilibrium of the system. At equilibrium, the demand allows the product to cost the optimum market price, known as the market clearing price '*Pe*', and the optimum quantity of the product '*Qe*'. Equilibrium in a market scenario is defined as "a situation in which plans of buyers and sellers exactly mesh, causing the quantity supplied to equal the quantity demanded at price in the market place for the good (product)" [128].

The Cournot model uses the supply and demand curve, where demand and quantities of the product determine the price. Figure 6.7 depicts a diagrammatic representation of the algorithm as a series of steps followed in the Cournot model. The model carries a number of assumptions:

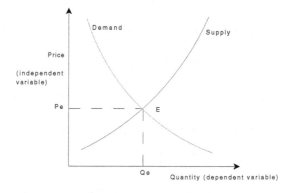

FIGURE 6.6: A supply-demand curve.

- All firms can be homogeneous or heterogenous in nature. Homogeneous means they have the same properties in memory like the same production costs, or heterogeneous with different costs.

- All produced goods are sold in the same iteration.

- The market price of the good is determined by the quantity of all other firms.

- The demand in the system can be static, the same throughout the simulation, or change dynamically during the simulation.

- The model exhibits an equilibrium which is a particular point at which the price is such that the quantity demanded is equal to the quantity supplied.

Figure 6.7 depicts a one iteration plan for the agents in the Cournot simulation. The figure also explains the various functions being performed during an iteration and how the price is determined.

The firms in a Cournot competition make decisions independent of other firms in the market. Their decisions can be improved by including learning in firms, which allows them to learn about their profits at the various outputs they produced. For instance, if producing high quantities causes high losses such as less sales, the firms compensate by producing less in the next time step.

With time, firms can individually find optimum quantity to produce at which they can all attain highest profits. This would be the equilibrium in the system, because firms will not have an incentive to move away from the equilibrium. This equilibrium is known as the Nash equilibrium or the Cournot-Nash equilibrium. At this point, none of the firms benefit by having different outputs. Sets of equations can be used to predict this behavior and the equilibrium

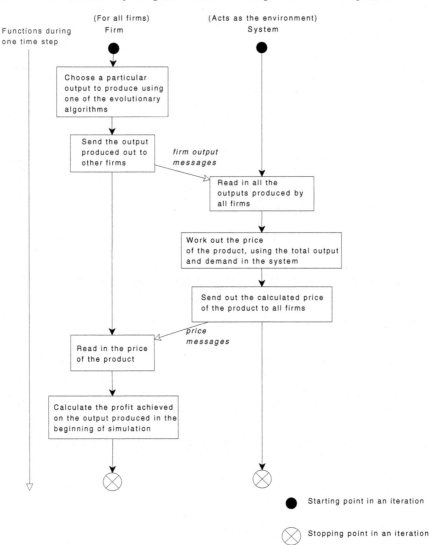

FIGURE 6.7: Time line of the various activities in a simple Cournot model.

point. For instance, the demand of a product is defined as

$$Demand = Q_{max} - Q \qquad (6.1)$$

where Q_{max} is the total demand of the product and Q is the total quantity produced by all firms at the time.

For simplicity all firms in the current experiment were assumed to be identical, allowing their cost functions to be equal. Each firm would thus have a cost of £10 for every good produced. Thus the cost function of the firm i can be given as

$$cost_i = 10 \times q_i \qquad (6.2)$$

where q_i is the quantity produced by firm i.

Therefore having established demand and quantities produced by the firms, the market price is given by

$$P_{market} = P_{zero} \times (Q_{max} - Q) \qquad (6.3)$$

where P_{zero} is the starting price of the product, assumed to be £1 in the experiment.

Given this price, the profit of each firm i can be calculated by

$$Profit_i = (P_{market} \times q_i) - (cost_i \times q_i) \qquad (6.4)$$

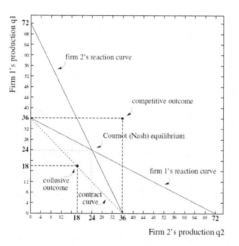

FIGURE 6.8: Firm reaction curves in a duopoly model. A duopoly market is a market with only two acting firms. Adapted from [5].

Using these mathematical notations (Equations 6.1-6.4) graphs can be plotted to show exactly where the equilibrium will occur. Figure 6.8 depicts the reaction curves of two firms competing in a duopoly market. If *Firm 1*'s production is zero, *Firm 2* can dominate the market by producing quantity equal to the demand. In this case, Firm 1 and 2, the demand in the system is 72. When *Firm 1* starts producing, *Firm 2* should reduce its output as the total quantity being produced becomes more than the demand. If there is too much of the product, this reduces its sales and *Firm 2* will suffer high loss. Note, that all products have to be sold in the same iteration. If the total quantity produced is more than the demand in the system, the product is given away for free or on negative market prices (Equation 6.3).

Figure 6.8 shows that Nash equilibrium occurs when both firms' reaction curves cross at values of $x = y = 24$. This is when both firms are producing the same output.

The Cournot model has been used as a basis to test patterns in firm behavior. Alkemade [5] depicted a similar design while writing an evolving Cournot model where the firms use genetic algorithm operations on the strategy base to optimize their productions as shown in Figure 6.9.

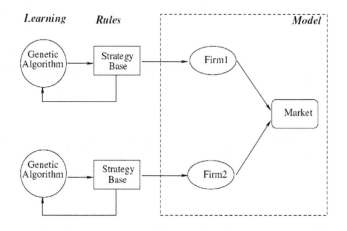

FIGURE 6.9: An evolutionary model. Each firm has its own strategy base which after every simulation is updated using genetic algorithms. Adapted from [5].

Alkemade used the experiment in a smaller duopoly market with only two firms, investigating use of evolutionary algorithms to study endogenous and exogenous factors which affected the Cournot equilibria. The study also concluded that simple agents performed well in static conditions, whereas more sophisticated agents with complex structures, like conditional or autoregressive agents, performed better in dynamic settings.

Vriend et al. [28] analyzed imitation behavior of firms, arguing that learning about the environment becomes more complex when there are too many choices. Their results showed that the firms are reluctant to imitate other firms, because they concluded that imitating a successful player would put them in a worse situation to begin with. Barr and Saraceno [19] also modeled a duopoly market framework focusing on how learning affected the equilibria of the system. The authors characterized firms as an artificial neural network, estimating outputs depending on signals received from the environment.

Arifovic used genetic algorithms to study adaptive behavior in firms in various economic models [9, 10]. In addition to using genetic operations, she also used an extra variable called election which helped compare results with rational expectations of firms. This was similar to the expected strategy used in the coevolutionary approach compared to actual played strategies. Arifovic [9] argued that the fluctuating behavior eventually converging to the equilibrium is not possible by standard approaches like least square methods, previously used in traditional economic theories. Dawid [47] supported the argument by saying that "genetic algorithm learning yielded qualitatively similar aggregate behavior than a population of human agents. The match is not perfect since the amplitude of oscillations decreases faster in genetic algorithms compared to the other approaches, such as least square learning method, these results are very satisfying."

Altavilla et al. [6] experimented with heterogeneous firms and compared the results to the Bertrand model. Friedman [70] supported the idea that players in reality behaved as if they have formulas in their heads. "It is only a short step from these examples to the economic hypothesis that under a wide range of circumstances individual firms behave as if they were seeking rationally to expected returns". Price [152] compared the evolution of price in Cournot and Bertrand models.

TABLE 6.2: Evolving Cournot characteristics for each firm.

Objective	Find the maximum profit that can be earned when competing with other firms.
Strategy representation	Quantity production represented as binary string of 9 bits can be converted into a numeric value.
Fitness case	Profit of the firm.
Selection scheme	Fitness proportionate roulette wheel selection.
Mutation rate	0.01, 0.03, 0.1
Crossover rate	0.1, 0.5, 0.7
Length of simulation	500
Number of runs averaged	20

The output the firms produce is represented as a string of binary digits. This allows the genetic operations, like crossover and mutation, to be performed easily on a numerical value. For example, $000100010 = (0 \times 2^8) + (0 \times 2^7) + (0 \times 2^6) + (1 \times 2^5) + (0 \times 2^4) + (0 \times 2^3) + (0 \times 2^2) + (1 \times 2^1) + (0 \times 2^0) = 34$.

The crossover and mutation rates help introduce variety in the population of strategies in the database. These can be introduced with different rates to allow divergence in the strategy population, at the same time preventing strategies from converging before all strategies have been tried.

Figure 6.10 depicts each firm having a strategy base which is maintained in the firm's memory. The strategy base looks like a database table with the

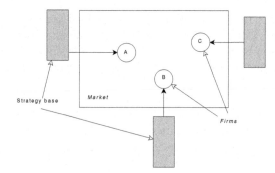

FIGURE 6.10: All firms have a strategy base in their memory.

binary string representing the strategy (production) and the profit acting as
the performance of the strategy (Figure 6.11).

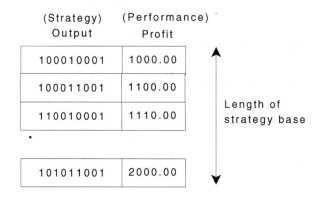

FIGURE 6.11: How a strategy base looks in firm's memory.

The strategy being evolved is represented as a string of bits to allow the
crossover and mutation techniques to be applied. The profit serves as the fitness or the result of applying different productions. This can also be called
the performance or payoff received for the strategy. As there is no optimization performed on the profit itself, it is not required to be represented as a
string. Thus using it as a double value variable serves the purpose of a payoff in the experiment. Figures 6.12 and 6.13 demonstrate how the crossover
and mutation functions will produce new strategies for the firms during the
simulation.

Three homogeneous firms were modeled in a system and the experiments
were run 20 times and an average collected. Figure 6.14 describes the Cournot
system with three firms and one system demand agent. The system demand
agent acts like the environment, collecting the firm outputs and calculating

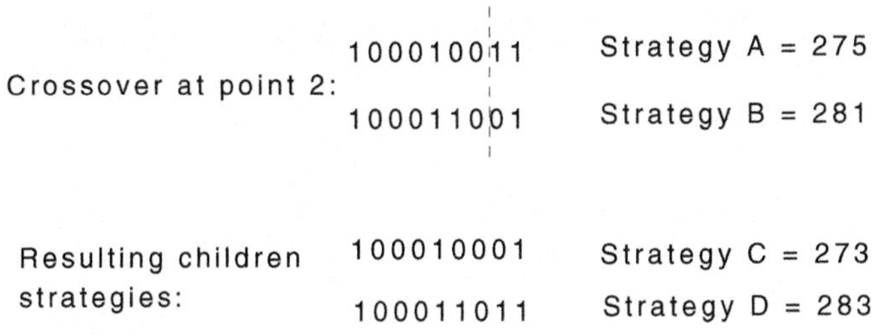

Crossover at point 2:

100010011   Strategy A = 275

100011001   Strategy B = 281

Resulting children
strategies:

100010001   Strategy C = 273

100011011   Strategy D = 283

FIGURE 6.12: How crossover works in Cournot model. Strategy A, B, C and D represent numerical values of bit strings.

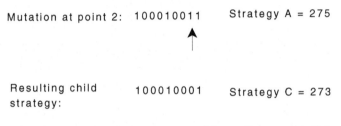

Mutation at point 2:   100010011   Strategy A = 275

Resulting child
strategy:

100010001   Strategy C = 273

FIGURE 6.13: How mutation works in Cournot model.

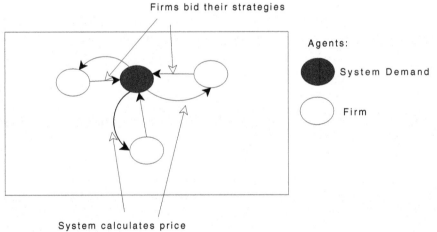

FIGURE 6.14: The system contains four agents (three firms and one system demand, responsible for assessing the product price).

the price in the system. Depending on the price, the firm agents can then calculate how much profit they have received for each bided strategy.

Using the equations discussed earlier (Equations 6.1 - 6.4) the Nash equilibrium was calculated theoretically to compare the experimental results. These have been listed in Table 6.3.

TABLE 6.3: Numerical values in Cournot experiment.

Variable	Value
$QMAX$	511 (Assuming all bits in a 9 digit binary string was 1)
n	3 (Number of firms)
Q^*	127.5 (Quantity at equilibrium)
P^*	128.5 (Price at equilibrium)
$Profit^*$	15108.75 (Profit at equilibrium)

The steps taken during the Cournot model are as follows:

Step 1: Firm Agent: If the beginning of the simulation, generate strategies for firm database; else do nothing.

Step 2: Firm Agent: Select an elitist strategy from the memory database based on roulette wheel selection. Post this as the chosen strategy.

Step 3: Demand Price Agent: Reads in all played strategies by firms and calculates the price of the product depending on the demand in the system.

Step 4: Firm Agent: Reads in the price of the product and calculates the actual profit as a result of playing the strategy.

Step 5: Firm Agent: Choose two elitist strategies from the database using new fitness. Perform crossover and mutation techniques to find three child strategies and save them.

Evolution relies on the trial and error process, trying best strategies and keeping a record of the most successful to produce new strategies. Figure 6.15 displays the results on quantities bid and profits collected. The quantities (Figure 6.15(a)) depict large variations as the simulation does not stabilize even after running it for 500 time steps. In Figure 6.15(b), profits were seen to converge close to the equilibrium value.

Between $t = 100$ and $t = 250$ (Figure 6.16(a)), the price comes very close to the ideal equilibrium price and oscillates about it, until at $t = 250$, one of the self-interested firms bids a higher quantity to attain higher profits. Deviation from the Nash equilibrium produces a loss to the firms, breaking the balance

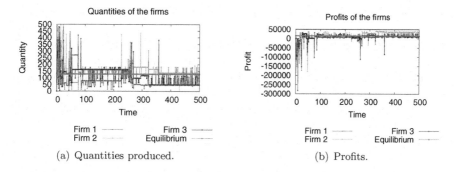

(a) Quantities produced.

(b) Profits.

FIGURE 6.15: Quantity and profits of three firms.

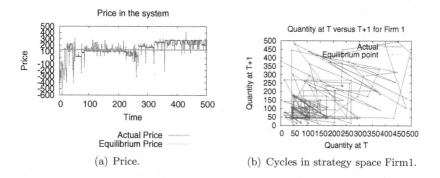

(a) Price.

(b) Cycles in strategy space Firm1.

FIGURE 6.16: Price and strategy space in evolution.

attained. Thus, firms try to come closer to equilibrium again which was later achieved after $t = 350$.

Figure 6.16(b) illustrates the plots of quantities produced at time t versus those produced at time $t = t + 1$. As all three firms were homogeneous, the strategy space of only one firm (*Firm* 1) is shown. Here the firm is trying out every possible strategy, making the strategy space spread across the graph. Although in the end it does concentrate around the equilibrium, the firm had to try a large number of possibilities before this was found.

Figures 6.17 and 6.18 depict the variation in prices, when crossover and mutation probabilities are varied. Price is a central variable, affected by quantities and profits of all firms. The simulations show that by increasing the crossover rates, the time for the price to find equilibrium increased. For instance, crossover rate 0.1 finds equilibrium at $t = 50$, 0.5 at $t = 100$ and 0.7 at $t = 200$.

Varying the mutation rate also shows similar behavior of extending time to oscillate about the equilibrium. This is very evident in $P_{crossover} = 0.7$ with the price stretched by increasing the mutation rates.

The model details are as follows:

Memory of Firm Agent. (int) id, (strategy) firm_strategy_map[10], (strategy) current_strategy, (strategy) chosen_strategy, (double) profit, (double) cost, (int) quantity, (double) avg_fitness.

List of Firm Functions. Firm_generate_strategies, Firm_idle, Firm_select _and_post_representative,   Firm_read_reps_evolve,   Firm_generate _childstrategy, Firm _play_strategy, Firm_collect_actual_fitness, Firm _post_histogram.

Memory of Demand Price Agent. (int) id, (double) pzero, (double) pt, (int) qmax.

List of Demand Price Functions. Demand_Price_calculate.

Memory of Averager Agent. (double) profit_firm1, (double) profit_firm2, (double) profit_firm3, (double)price, (double) quantity_firm1, (double) quantity_firm2, (double) quantity_firm3.

List of Averager Functions. Averager_collect.

Messages. Current_strategy: Contains firm_id, production, scene_id; priceP: Contains mprice, scene_id; current_profit: Contains firm_id, quantity, profit, scene_id; firm_representative: Contains firm_id, current_strategy, scene_id; my_chosen_strategy: Contains firm_id, chosen_quantity, scene_id

Datatypes. Strategy: Contains output[9], score, frequency

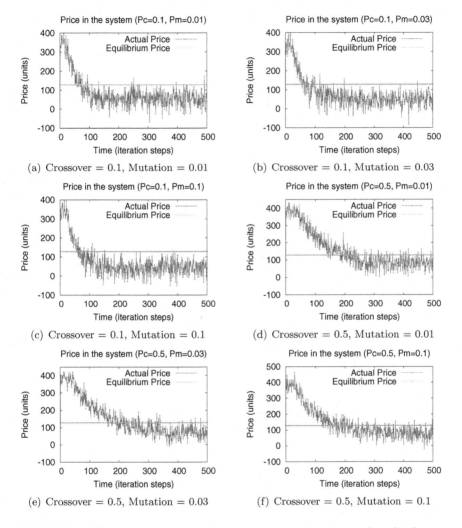

FIGURE 6.17: Average price with crossover rate 0.1, 0.5 and multiple mutation rates.

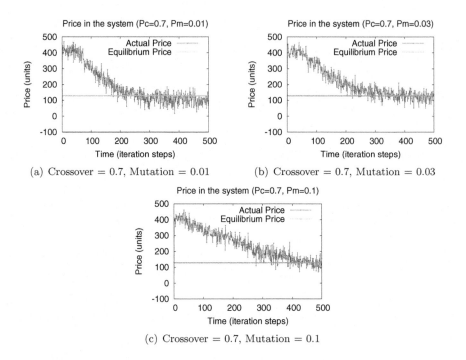

(a) Crossover = 0.7, Mutation = 0.01   (b) Crossover = 0.7, Mutation = 0.03

(c) Crossover = 0.7, Mutation = 0.1

FIGURE 6.18: Average price with crossover rate 0.7 and multiple mutation rates.

6.3.1   Genetic Programming with Agents

```
<!-- extract from model XML file-->
<dataTypes>
  <dataType>
  <name>strategy</name>
  <variables>
    <variable><type>int</type><name>output[9]</name></variable>
    <variable><type>double</type><name>score</name></variable>
    <variable><type>int</type><name>frequency</name></variable>
  </variables>
</dataType>

......

<xagent>
<name>Firm</name>
<memory>
  <variable><type>int</type><name>id</name></variable>
  <variable><type>strategy</type><name>firm_strategy_map[5]</name></variable>
  <variable><type>strategy</type><name>current_strategy</name></variable>
  <variable><type>int</type><name>x</name></variable>
  <variable><type>int</type><name>y</name></variable>
  <variable><type>double</type><name>profit</name></variable>
  <variable><type>double</type><name>cost</name></variable>
  <variable><type>int</type><name>quantity</name></variable>
  <variable><type>double</type><name>avg_fitness_opt</name></variable>
  <variable><type>double</type><name>avg_fitness</name></variable>
</memory>

<functions>
  <function>
  <name>Firm_select_strategy</name>
  <currentState>00</currentState>
  <nextState>00a</nextState>
  <inputs>
    <input><messageName>strategies_for_firm</messageName>
    <filter>
      <lhs><value>a.id</value></lhs>
      <op>EQ</op>
      <rhs><value>m.firm_id</value></rhs>
    </filter>
    </input>
  </inputs>
  </function>

  <function>
  <name>Firm_play</name>
  <currentState>00a</currentState>
  <nextState>01</nextState>
  <outputs>
    <output><messageName>current_strategy</messageName></output>
  </outputs>
  </function>

  <function>
  <name>Firm_calculate_profit</name>
```

```
  <currentState>01</currentState>
  <nextState>02</nextState>
  <inputs>
    <input><messageName>priceP</messageName></input>
  </inputs>
  <outputs>
    <output><messageName>current_profit</messageName></output>
  </outputs>
  </function>

  <function>
  <name>Firm_optimise</name>
  <currentState>02</currentState>
  <nextState>03</nextState>
  </function>
</functions>
</xagent>

/* The Firm Agent Functions**/

#include "header.h"
#include "my_library_header.h"
#include "Firm_agent_header.h"
#include <math.h>

int Firm_select_strategy()
{
  int counter=0; //to count empty strategies
  int counter_j=0;//to count inside strategy
  strategy_array temp_good_strategies;
  init_strategy_array(&temp_good_strategies);

  strategy_array temp_cum_fitness;
  init_strategy_array(&temp_cum_fitness);

  strategy_array new_strategies;
  init_strategy_array(&new_strategies);
  int new=0;
  double lowest=0.0;
  int pos=0;
  int temp_size=0;
  double sum=0.0, sum_prob=0.0;
  double start_ptr=0.0;
  int i,j,l;
  int strategy_chosen_flag=0;
  strategy received_strategies[5];
  int FirmStrategySize=10;

  //choose one of the good strategies in my database
  //using roulette wheel selection

  for(i=0;i<FirmStrategySize;i++)
  {
    if(FIRM_STRATEGY_MAP[i].score>0.0)
    {
      //only consider strategies which have more than 0 scores
      add_strategy(&temp_good_strategies,&FIRM_STRATEGY_MAP[i].output[0],
```

```
                        FIRM_STRATEGY_MAP[i].score,0);
   }
}
//output all chosen strategies
for(i=0;i<temp_good_strategies.size;i++)
{
   for(j=0;j<LENGTH;j++)
   {
     printf("%d,", temp_good_strategies.array[i].output[j]);
   }
   printf(" score %f\n",temp_good_strategies.array[i].score);
}

//choosing one of these strategies to play

sum=0.0;
sum_prob=0.0;
if(temp_good_strategies.size>0)
{
   for(i=0;i<temp_good_strategies.size;i++)
   {
     sum+=temp_good_strategies.array[i].score;//sum of fitness
   }

   for(i=0;i<temp_good_strategies.size;i++)
   {
     sum_prob+=(temp_good_strategies.array[i].score/sum); //sum of prob
     add_strategy(&temp_cum_fitness,
                 temp_good_strategies.array[i].output,sum_prob,0);
   }

   start_ptr=(double)rand()/(double)RAND_MAX;

   for(l=0;l<LENGTH;l++)
   {
     CURRENT_STRATEGY.output[l]=temp_good_strategies.array[0].output[l];
   }
   CURRENT_STRATEGY.score=temp_good_strategies.array[0].score;

   for(j=0;j<temp_good_strategies.size;j++)
   {
     if((start_ptr>=temp_cum_fitness.array[j].score)&&
                 (start_ptr<=temp_cum_fitness.array[j+1].score))
     {
       for(l=0;l<LENGTH;l++)
       {
         CURRENT_STRATEGY.output[l]=temp_good_strategies.array[j].output[l];
       }
       CURRENT_STRATEGY.score=temp_good_strategies.array[j].score;
     }//end if
   }//end for
}//end if
return 0;
}

int Firm_play()
{
```

```
  int i, j=0;
  double total=0.0;

  for(i=0;i<LENGTH;i++)
  {
    j=LENGTH-i-1;
    total=total+ (pow(2,j)*CURRENT_STRATEGY.output[i]);
  }
  QUANTITY=total;
  add_current_strategy_message(ID,total);
  return 0;
}

int Firm_calculate_profit()
{
  double price=0.0;
  double total_fitness=0.0;
  int i;
  //get profits from the Demand Price agent
  priceP_message=get_first_priceP_message();
  while(priceP_message)
  {
    price=priceP_message->mprice;
    priceP_message=NULL; //Stop the message loop once value found
  }
  PROFIT=(price-COST)*QUANTITY;
  CURRENT_STRATEGY.score=PROFIT;

  add_current_profit_message(ID,CURRENT_STRATEGY);

  //add score to current strategy database
  for(i=0;i<FirmBoardSize;i++)
  {
    if(compare_arrays(&FIRM_STRATEGY_MAP[i].output,
                        &CURRENT_STRATEGY.output)==0)
    {
      FIRM_STRATEGY_MAP[i].score=CURRENT_STRATEGY.score;
    }
  }

  for(i=0;i<FirmBoardSize;i++)
  {
    total_fitness+=FIRM_STRATEGY_MAP[i].score;
  }
  AVG_FITNESS=total_fitness/FirmBoardSize;

  return 0;
}
// Performs crossover and mutation
int Firm_optimise()
{
  int i=0;
  double total_fitness=0.0;
  strategy parent;
  init_strategy(&parent);
  strategy_array temp_good_strategies;
  init_strategy_array(&temp_good_strategies);
```

```
strategy_array temp_cum_fitness;
init_strategy_array(&temp_cum_fitness);
int j,l;
double sum=0.0;
double sum_prob=0.0;
double start_ptr;
double do_cross=0.0;
double do_mutate=0.0;
int crossover_point=0;

int mutation_point=0;
int pos=0;
double lowest=0.0;
strategy childOne;
strategy childTwo;
strategy mutantChild;

init_strategy(&childOne);
init_strategy(&childTwo);
init_strategy(&mutantChild);
int newchildOne=0,newchildTwo=0;
int newmutantChild=0;

int found=0;

if(CURRENT_STRATEGY.score>=THRESHOLD)
{
  //select two parents via Roulette Wheel Selection
  for(i=0;i<FirmBoardSize;i++)
  {
    if(FIRM_STRATEGY_MAP[i].score>0.0)
    {
      //prevent choosing the same parent
      if(compare_arrays(&FIRM_STRATEGY_MAP[i].output,
              &CURRENT_STRATEGY.output)==1)//false
      {
        add_strategy(&temp_good_strategies,
                        &FIRM_STRATEGY_MAP[i].output[0],
                        FIRM_STRATEGY_MAP[i].score,0);
      }
    }
  }
  for(i=0;i<temp_good_strategies.size;i++)
  {
    sum+=temp_good_strategies.array[i].score;//sum of fitness
  }
  for(i=0;i<temp_good_strategies.size;i++)
  {
    sum_prob+=(temp_good_strategies.array[i].score/sum); //sum of prob
    add_strategy(&temp_cum_fitness,
                    temp_good_strategies.array[i].output,sum_prob,0);
  }
  start_ptr=(double)rand()/(double)RAND_MAX;
  for(j=0;j<temp_good_strategies.size;j++)
  {
    if(start_ptr<temp_cum_fitness.array[0].score)
```

```
      {
        for(l=0;l<LENGTH;l++)
        {
          parent.output[l]=temp_good_strategies.array[0].output[l];
        }
        parent.score=temp_good_strategies.array[0].score;
      }
      if((start_ptr>=temp_cum_fitness.array[j].score)&&
                    (start_ptr<=temp_cum_fitness.array[j+1].score))
      {
        for(l=0;l<LENGTH;l++)
        {
          parent.output[l]=temp_good_strategies.array[j].output[l];
        }
        parent.score=temp_good_strategies.array[j].score;
      }
    }

    //performing crossover
    do_cross=(double)rand()/(double)RAND_MAX;
    if(do_cross<=CROSSOVER_RATE)
    {
      crossover_point=rand()%LENGTH;
      for(i=0;i<crossover_point;i++)
      {
        childOne.output[i]=CURRENT_STRATEGY.output[i];
        childTwo.output[i]=parent.output[i];
      }

      for(i=crossover_point;i<LENGTH;i++)
      {
        childOne.output[i]=parent.output[i];
        childTwo.output[i]=CURRENT_STRATEGY.output[i];
      }
      newchildOne=0;
      newchildTwo=0;
      for(i=0;i<FirmBoardSize;i++)
      {
        if(compare_arrays(&childOne.output,
            &FIRM_STRATEGY_MAP[i].output)==0)
        {
          newchildOne=1;
        }
        if(compare_arrays(&childTwo.output,
                  &FIRM_STRATEGY_MAP[i].output)==0)
        {
          newchildTwo=1;
        }
      }

//produce offspring through crossover & mutation
//replace lowest fitness with new offsprings
if(newchildOne==0)
{
  for(i=0;i<FirmBoardSize;i++)
      {
        lowest=FIRM_STRATEGY_MAP[0].score;
```

```
              pos=0;
              for(j=0;j<FirmBoardSize;j++)
              {
                if(lowest>FIRM_STRATEGY_MAP[j].score)
                {
                  lowest=FIRM_STRATEGY_MAP[j].score;
                  pos=j;
                }
              }

      for(l=0;l<LENGTH;l++)
      {
              FIRM_STRATEGY_MAP[pos].output[l]=childOne.output[l];
      }
      FIRM_STRATEGY_MAP[pos].score=THRESHOLD;
}
        }
      //second child
      if(newchildTwo==0)
      {
              for(i=0;i<FirmBoardSize;i++)
              {
      lowest=FIRM_STRATEGY_MAP[0].score;
      pos=0;
      for(j=0;j<FirmBoardSize;j++)
      {
        if(lowest>FIRM_STRATEGY_MAP[j].score)
                {
                  lowest=FIRM_STRATEGY_MAP[j].score;
                  pos=j;
                }
      }
      for(l=0;l<LENGTH;l++)
      {
        FIRM_STRATEGY_MAP[pos].output[l]=childTwo.output[l];
      }
      FIRM_STRATEGY_MAP[pos].score=THRESHOLD;
          }
        }
      }//end of do_cross loop
      if(do_mutate<=MUTATION_RATE)
      {
        mutation_point=rand()%LENGTH;
        for(l=0;l<LENGTH;l++)
        {
          mutantChild.output[l]=CURRENT_STRATEGY.output[l];
        }
        mutantChild.score=THRESHOLD;
        if(mutantChild.output[mutation_point]==1)
        {
          mutantChild.output[mutation_point]=0;
        }
        else if(mutantChild.output[mutation_point]==0)
        {
          mutantChild.output[mutation_point]=1;
        }
        else{}
```

```
    newmutantChild=0;
    for(i=0;i<FirmBoardSize;i++)
    {
      if(compare_arrays(&FIRM_STRATEGY_MAP[i].output,
                        &mutantChild.output)==0)//true
      {
        newmutantChild=1;
      }
    }

    if(newmutantChild==0)
    {
      for(i=0;i<FirmBoardSize;i++)
      {
        lowest=FIRM_STRATEGY_MAP[0].score;
        pos=0;
        for(j=0;j<FirmBoardSize;j++)
        {
          if(lowest>FIRM_STRATEGY_MAP[j].score)
          {
            lowest=FIRM_STRATEGY_MAP[j].score;
            pos=j;
          }
        }
        for(l=0;l<LENGTH;l++)
        {
          FIRM_STRATEGY_MAP[pos].output[l]=mutantChild.output[l];
        }
        FIRM_STRATEGY_MAP[pos].score=THRESHOLD;
      }
    }
  }//end if mutate
}//end if(CURRENT_STRATEGY.score>=THRESHOLD)
for(i=0;i<FirmBoardSize;i++)
{
  total_fitness+=FIRM_STRATEGY_MAP[i].score;
}
AVG_FITNESS_OPT=total_fitness/FirmBoardSize;
return 0;
}
```

6.3.2   Filtering Messages in Advance

Messages can be filtered in advance at message boards, before functions start reading them. This is handled by FLAME if defined in the XML description file. The C function stays exactly the same, but speeds up simulation time in cases where a large number of messages are read to find particular values.

```
<function>
  <name>Firm_select_strategy</name><description></description>
  <currentState>00</currentState>
  <nextState>00a</nextState>
  <inputs>
```

```
<input><messageName>strategies_for_firm</messageName>
<filter>
  <lhs><value>a.id</value></lhs>
  <op>EQ</op>
  <rhs><value>m.firm_id</value></rhs>
</filter>
</input>
</inputs>
</function>

/* Message reading stay exactly the same*/
strategies_for_firm_message=get_first_strategies_for_firm_message();
while(strategies_for_firm_message)
{
  if(strategies_for_firm_message->firm_id==ID)
  {
    for(i=0;i<FirmBoardSize;i++)
    {
      for(j=0;j<LENGTH;j++)
      {
        received_strategies[i].output[j]=
                      strategies_for_firm_message->
                      firm_strategies[i].output[j];
      }
      received_strategies[i].score=strategies_for_firm_message->
                                  firm_strategies[i].score;
    }
  }
}
strategies_for_firm_message=get_next_strategies_for_firm_message
                                  (strategies_for_firm_message);
}
```

6.3.3 Comparing Two Data Structures

Two arrays or datatypes can be compared, to see if they carry the same values or not. This is useful when performing crossover between two parents, to check if both parents are not the same strategy strings. By using the 'compare_array' function, if a zero if returned this is true; else 1 means a false.

```
if(compare_arrays(&received_strategies[i].output,
                  &FIRM_STRATEGY_MAP[j].output)==0)
{
  //true they are the same
  ...//if 1 is returned, they are two different strings
}
```

6.4   A Virtual Mall Model: Labor and Goods Market Combined

The virtual mall model is a simple representation of interactions between the labor and goods markets. The model uses learning to allow agents to learn their most profitable strategies in a changing environment. Learning allows new strategies to be produced in the market, which are not pre-coded at the start of the experiment. The model involves four agents described below.

Malls. The Mall agents functions on a monthly cycle, at every 20 workable days. At the start of every month, the malls use a strategy posted to them by the environment agent and use it for 5 days. At the end of 5 days they assess its performance. The current performance is compared to the gradient at the end of the previous month, at which point the mall decides to adapt this new strategy for the rest of the month or switch to the old one.

Persons. The Person agents possess an extra variable in their memory called *the learning window*. If the person strategy proves to be better, the learning window increases in length, causing the assessment period for the person to appear at a later stage. Thus the lengths of the windows can give us insights on whether the people were performing well or not as well as their gradients.

Environment. This agent is responsible communicating information to the agents. The strategies are also held in the environment and are posted at the time the agents need to try out new strategies. This is in contrast for keeping the strategies in the agent memories. The environment also holds the messages for the agents which are to be used within the strategy. By keeping them here, we are able to remove the communication dependency within the strategy gene of every agent allowing free access for functions to be moved around.

Message counter. This agent is responsible for keeping track of the messages being sent between agents. This would allow us to see the reasons for some of the emerging behavior of agents.

The experiment was used to test various hypotheses and understand how learning and behavior are seen in a labor and goods market model. The simulation was started with neither the malls nor the people having knowledge about their previous actions and the results were as follows.

Best performing malls. The malls were told to go bankrupt if their capitals went below a bankruptcy level. Therefore, whenever this happened, the malls were removed from the simulation. The list of functions which the malls could include in their strategies were

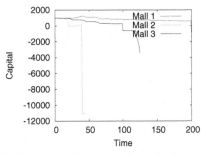
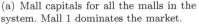

(a) Mall capitals for all the malls in the system. Mall 1 dominates the market.

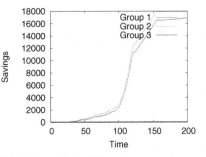
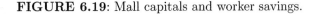

(b) Savings of the different groups of people.

FIGURE 6.19: Mall capitals and worker savings.

1. Advertise.
2. Do nothing.
3. Hire workers.
4. Sell goods.
5. Make redundant.
6. Assess goods price.
7. Assess goods productions.
8. Give promotions.

The mall strategy gene is of length seven, as it allowed to perform seven functions in one iteration. Figure 6.19(a) shows that Mall 1 dominates the market. Mall 3 tries to compete but is unable to catch up to the prior mall and Mall 2 goes bankrupt at time $t \approx 50$. Mall 1 benefits from the start by using the following gene of functions,

advertise-do nothing-promotions-sell goods-assess goods price-assess production

Employing this strategy, keeps the mall capital constant and rising. At periods $t2, t3$, the mall hires people for work. By periodically hiring people and selling goods, the mall steadily rises its capital. However, this strategy is not good for a long time, where the mall switches to another gene of functions, which consists of varying goods production depending on sales and advertising.

Mall 2 starts by employing function in its gene, which is

do nothing-promotions-hire worker-hire worker-make redundant-sell goods-assess production

Here the mall tries to hire more workers in its strategy, we see a more sharper fall in capitals. Although the mall switches to selling as the Mall 1 did, it cannot recover most of its losses, and goes bankrupt sharply.

Interestingly, also self-evolving Mall 3 started with Mall 2's behavior, but did not adopt it for long. This is probably due to people being employed by Mall 2 and no more applications were left. Also there are not many sales for Mall 2. Mall 2 runs before in the agent order during the simulation. Thus Mall 3 tries out new strategies but keeps closing down. Because of so many inactive periods, the mall in unable to climb up in its sales and soon goes bankrupt.

Best performing people groups. There were three groups of people which have the same information within the group. It was up to the individuals to use information. The gene length of people functions was five and included:

1. Find job.
2. Buy cheap goods.
3. Buy advertised.
4. Buy recommended.
5. Buy random.
6. Do nothing.
7. Quit job.

Figure 6.19(b) shows that Group 1 and 2 are able to compete with each other, whereas Group 3 can not. Savings are largely influenced by wages of the people. Figure 6.20(a) shows the different wages of the people. As the wages of people in Group 2 are higher, they are able to save up more money than Group 1.

Since Mall 2 goes bankrupt at an early stage, it does not show any impact on people savings. But when Mall 3 goes bankrupt at $t \approx 125$, we see a considerable fall in the gradient of savings. The demand wage decreases as more and more people become jobless and try to get new jobs with other existing malls.

Learning windows of the people. In Figures 6.19(b) and 6.20(a), Group 2 is doing better than Group 1. This is due to their learning windows being smaller, so people in Group 2 are quicker to react and change behavior and adapt better.

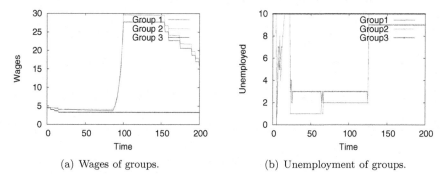

(a) Wages of groups.   (b) Unemployment of groups.

FIGURE 6.20: People wages and unemployment.

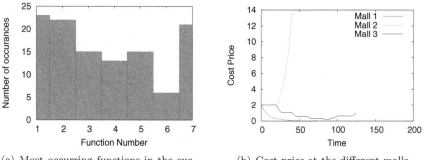

(a) Most occurring functions in the suc-   (b) Cost price at the different malls.
cessful genes.

FIGURE 6.21: Mall strategies and costs.

Figure 6.20(b) shows that Group 3 never gets employed and they never do well in the system. When the mall goes bankrupt at t=125, there are large numbers of unemployed people in the system with Group 2 completely out of jobs. Thus Group 2 fails at 200 iterations.

Functions used by the groups. Because every person updates their strategies at different times it is difficult to say which are the most successful strategy genes used. Figure 6.21(a) shows that groups recognize that it is more profitable to have functions like looking for jobs and buying cheaper goods than the rest. Buying goods-advertised, recommended or random are seen in lower numbers in the strategies used.

Cost price of the malls. Figure 6.21(b) shows that Mall 2 increased the cost price too much leading to fewer sales. Mall 1, selling goods at cheap prices, benefitted in the long run.

Model details are as follows, with Figure 6.22 showing the model state graph:

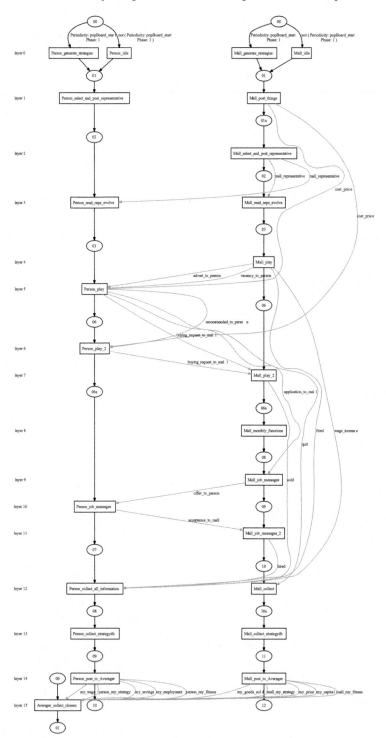

FIGURE 6.22: Stategraph for virtual mall experiment.

- Memory of Mall Agent. (int) id, (mall_strategy) current_mall_function_map, (mall_strategy) previous_mall_function_map, (double) previous_mall_profit, (int) retailer[25], (int) security, (double) current_mall_profit, (int) stock, (int)stock_sold, (int) stock_addition, (double) cost_price, (double) retailer_wage[25], (double) capital, (int) region_id, (int) x, (int) y, (double) capitalPminus1, (double) capitalP, (double) previous_capital_gradient, (double) current_capital_gradient.

- Memory of Person Agent. (int) id, (person_strategy) current_person_function_map, (person_strategy) previous_person_function_map, (double) previous_performance, (double) current_performance, (double) savings, (int) employer_mall_id, (double) expenses, (int) goods_bought, (double) wage, (int) region_id, (int) x, (int) y, (double) savingsPminus1, (double) savingsP, (double) previous_savings_gradient, (double) current_savings_gradient, (double) performance_gradient, (int) learning_window, (int) learning_window_assess.

- Memory of Environment Agent. (int) id, (mall_strategy_array) mall_actions, (person_strategy_array) person_actions, (d_vacancy_message_array) vacancy_msg, (d_application_message_array) application_msg, (d_offer_message_array) offer_msg, (d_acceptance_ message_array) acceptance_msg, (d_advert_message_array) advert_msg, (d_buying_request_message_array) buying_request_msg, (d_recommended_message_ array) recommended_msg.

- List of Mall Functions:

  - Mall_generate_strategies
  - Mall_idle
  - Mall_post_things
  - Mall_select_and_post_representative
  - Mall_read_reps_evolve
  - Mall_play
  - Mall_play_2
  - Mall_monthly_functions
  - Mall_job_messages
  - Mall_job_messages_2
  - Mall_collect
  - Mall_collect_strategydb
  - Mall_post_to_Averager

- List of Person Functions:

- Person_generate_strategies
- Person_idle
- Person_select_and_post_representative
- Person_read_reps_evolve
- Person_play
- Person_play_2
- Person_job_messages
- Person_collect_all_information
- Person_collect_strategydb
- Person_post_to_Averager

- Messages:

    - mall_representative: Contains mall_id, current_strategy, scene_id
    - advert_to_person: Contains mall_id, scene_id
    - vacancy_to_person: Contains mall_id, shop_id, scene_id
    - buying_request_to_mall: Contains person_id, mall_id, costprice, scene_id
    - sold: Contains person_id, mall_id, cost_price, scene_id
    - fired: Contains person_id, compensation, scene_id
    - hired: Contains person_id, mall_id, wage, scene_id
    - recommended_to_person: Contains mall_id, scene_id
    - wage_increase: Contains mall_id, person_id, wage, scene_id
    - cost_price: Contains mall_id, price, scene_id
    - application_to_mall: Contains person_id, mall_id, wage, scene_id
    - quit: Contains person_id, mall_id, scene_id
    - offer_to_person: Contains person_id, mall_id, wage, scene_id
    - acceptance_to_mall: Contains person_id, mall_id, wage, scene_id
    - person_my_fitness: Contains person_id, score, predicted_score, scene_id
    - my_employment: Contains person_id, mall_id, scene_id
    - my_savings: Contains person_id, savings, scene_id
    - person_my_strategy: Contains person_id, current_strategy, scene_id
    - my_wage: Contains person_id, wage, scene_id
    - mall_my_fitness: Contains mall_id, score, predicted_score, scene_id
    - my_capital: Contains mall_id, capital, scene_id

 - mall_my_strategy: Contains mall_id, current_strategy, scene_id
 - my_goods_sold: Contains mall_id, goods, scene_id
 - my_price: Contains mall_id, price, scene_id

- Datatypes:

 - mall_strategy: Contains output[7], score, predicted_score
 - person_strategy: Contains output[5], score, predicted_score

---

6.5 Programming Games

Neumann and Morgenstern's work [199] pioneered interdisciplinary research in game theory, using concepts of expected utility to explain a person's 'betting preferences' with regard to uncertain outcomes in a gaming situation. The authors also described the concept of rationality by comparing economic situations with the *Robinson Crusoe* model, where the system's complete economy is led by one individual who is responsible for all rules imposed in a closed isolated system. The objective of the individual is to perform tasks and impose rules which would eventually maximize their own benefit. However, the model ignored factors such as weather, savages or crops which would eventually influence the decisions made. Weighing these factors can be introduced through probabilities of their influence on economic decision outcomes.

Departing from the idea of a single individual is the concept of *social economy*, which involves more than one individual interacting with others, in turn presenting a different sets of challenges to the economy. The social interaction provides individuals with more or limited information, through their networks, who can then make decisions based on this for their own benefit. The strategies used in each situation, and by each individual, are different, working to find a maxima for the individual performing in the situation. This maxima or maximum value represents the utility or the performance of the variable being optimized through the strategies. With this argument, each individual would behave rationally to maximize their utility and choose the most optimum strategy in the situation. However, in reality, recent work has argued the influence of cognitive psychology, bias and chance on rational decisions in economic scenarios such as in the works of McFadden [130] and Kahnemann [99].

"Game theory has developed powerful tools for analyzing decision making in systems with multiple autonomous actors. These tools, when tailored to computational settings, provide a foundation for building multi-agent software systems. This tailoring gives rise to the field of computational mechanism design, which applies economic principles to computer systems designs." [46]

Games have been used with economics to explain how individual players adopt different strategies when trying to constantly outsmart each other [157]. Game theory embodies research as different kinds of games and is essentially the study of these strategies. Most games have payoff matrices that determine the profit received by the agent when a certain strategy is played. Economists widely used game theoretic approaches to model goal-directed behavior in agents as a way to emulate competitive and collaborative characteristics in humans.

The utility functions are embeded within players and allow them to assess their behavior. Each player is a self-interested individual, trying to improve their behavior by measuring it, using the utility function. However, this approach is still very limiting assuming all agents behave in predefined ways, ignoring the varied personalities of humans and other events affecting their decisions.

Since traditional economic ideologies are based on rational theory, game theory provides a number of advantages for scientists to view economic and social systems as game scenarios. These systems contain the following attributes:

- Introduce a rational choice theory for all players.

- The provision of the utility function which is maximized by all players.

- Investigate the concepts of domination using the Nash equilibrium.

6.5.1   Nash Equilibrium

Equilibrium in economics is another important concept, first projected in Walrasian models, where Walras was convinced that economics could be made predictable. He was influenced by the physics principles and imported the concept of equilibrium, laying mathematical foundations in traditional economics [21]. The theory supports the claim of the invisible hand, stating that whatever happens in the market, it would eventually reach an equilibrium which is the best scenario for all players or actors in the system.

In game theory, Nash equilibrium was proposed by John Forbes Nash [139, 138] in games that involve two or more players. Each player assumes to play the strategy that lies close to the equilibrium, which is the point when no player would benefit if it strayed from the current equilibrium strategies, being the best for all players.

Let (S, f) be a game of n players where S_i is the strategy set for player i. Thus the set of strategy profiles (Equation 6.5) would have associated payoff functions (Equation 6.6),

$$S = S_1 \times S_2 ... \times S_n \tag{6.5}$$

$$f = f_1(x), f_2(x)..., f_n(x) \tag{6.6}$$

Each player i would then choose a strategy x and obtain a certain payoff $f_i(x)$. In a Nash equilibrium, players will choose a certain strategy x^* such that no player would get a profit if they deviated from this strategy.

$$\forall i, x_i \in S_i, x_i \neq x^* : f_i(x_i^*, x_{-i}^*) \geq f_i(x_i, x_{-i}^*) \tag{6.7}$$

Games can have pure or mixed strategies for the players, affecting the Nash equilibrium reached by the players. Pure strategies are a set of strategies given to the player with details on when to apply them. In mixed-strategy games, players have a probability of choosing different strategies from their strategy set. In these situations, equilibrium is defined as the *trembling hand equilibrium* because there is a probability between the strategy choice. Some mixed-strategy games allow players to have more than one Nash equilibrium.

6.5.2 Evolutionary Game Theory

Recent interest of economists and biologists has moved from traditional game theory to *evolutionary game theory* as it provides more insights and analysis of systems, particularly reducing the number of assumptions.

Maynard Smith [186] extended the principle of classical game theory by applying it to a population dynamical setting. This work focused on the self-regulation within actual species who are competing together. By introducing self-regulation using the frequency of the species characteristics, the theory allows dynamic systems to be expressed mathematically. In his unpublished thesis work, Smith wrote "it is unnecessary to assume that the participants have ... the ability to go through any complex reasoning processes. But the participants are supposed to accumulate empirical information on the various pure strategies at their disposal... We assume that there is a population ... of participants... and that there is a stable average frequency with which a pure strategy is employed by the *average member* of the appropriate population" [85]. This work was largely based on principles of ecology.

Smith and Price [188] showed how animals adapted themselves to cope better with scenarios like territory domination and competing for mates. The authors presented the Hawk-Dove game where the players had no knowledge about the optimal strategies. Through Darwinian selection, the hawks and doves were able to evolve to an *evolutionary stable state* (ESS), which was the Nash equilibrium where the populations stabilized. Smith concluded that Darwinian selection could be substituted for agent rationality where the fitness of the strategies is determined by the survival of the player in the population.

6.5.3   Evolutionary Stable State

An important concept here is the ESSs, which differs from the strict definition of the Nash equilibrium (Table 6.4). ESSs study the strategies adopted by the players on a population level. This involves studying a large number of players. The ESS is achieved through the frequency of the players in the population. Thus ESS is a frequency-dependent concept which has been propagated by the selection mechanisms of evolution [186]. ESS allows for a given set of behaviors (conserved over time) to determine an optimal strategy for everyone in the system. At this time no other mutant behavior can enter the system and survive. This means that behavior is adopted by the individuals in the population, and no other behavior will invade the population under natural selection. Suppose the main population plays strategy $x \in S$ and mutants can play some strategy $y \in S$. Given that the mutants in the population are a very small proportion, then the probability a mutant is drawn for the population is very small probability ϵ by evolutionary selection. The payoff function for the strategies is determined by $u(x)$. An evolutionary stable state occurs when no mutant population can invade the main population. This can be true in the two conditions (Equation 6.8 and 6.9).

$$u(y, x) \leq u(x, x) \tag{6.8}$$

$$u(y, x) = u(x, x) \Rightarrow u(y, y) < u(x, y) \forall y \neq x \tag{6.9}$$

The concepts of ESS favor the analysis of dynamical systems which is why it is extensively used in biological systems [45].

6.5.4   Game Theory versus Evolutionary Game Theory

An important advantage of using ESS compared to Nash equilibrium is that Nash equilibrium can only be achieved using rational decisions and discrete payoffs in a game, whereas ESS does not depend on rational decisions. It is rather based on the behavioral aspects of the individuals. Despite this difference, there are some games of an altruistic nature in which the two definitions can be related. Prisoner's dilemma game is an example in which only when all players cooperate a Nash equilibrium is reached, benefitting all players in the game. These games use rational or discrete utilities as payoffs.

Silverberg [178] favors the use of replicator dynamics to model economic evolution. Replicator dynamics uses frequency-dependent fitness to depict the most used strategy in an economic scenario. Frequency-dependent fitness is different from using a payoff function for the fitness because the payoff function assesses the performance of a strategy and the strategy fitness is based on the performance rather than its frequency in the population.

6.5.5 Continuous Strategies

Most theories in game theory and evolutionary game theory are based on discrete strategies. Contrastingly in economics most strategies are continuous variables like the prices or the production by a company. In such situations the population stabilizes in one of the three following conditions [187]:

- There is a unique x, such that if all players play x, no mutant (new) y can invade the population.

- There is a certain distribution (δx), which states that population lies between (x) and $(\delta x + x)$.

- There is no ESS in the population.

6.5.6 Red Queen and Equilibrium

Red Queen Dynamics, a name borrowed from one of Lewis Caroll's [38] characters, is the term given to the constant evolutionary arms race between more than one species evolving together. For instance in the scenario of a predator-prey model, the predator is continuously searching for the prey and adapting its path to increase the chances of finding food. At the same time, the prey would be adapting itself to find new paths to get away from the predator to prevent being eaten. When put together both predator and prey enter into the *evolutionary arms race*, each working for their own benefit, evolving their behavior constantly.

Valen [196] describes the need for the Red Queen effect saying that "for an evolutionary system, continuing development is needed just in order to maintain its fitness relative to the systems it is coevolving with". However the Peter principle states "evolution systems tend to develop to the limit of their adaptive competence" [147]. This shows that at a certain point the system reaches a maximum, where their adaptiveness is not benefitting the actors any more. This stage can be referred to as the reaching of an equilibrium, where evolving further behavior by any of the species will not benefit either of them. This concept is similar to the *Nash equilibrium* and the solutions of Evolutionary Stable States as the populations momentarily stabilize at this point.

Researchers like Malthus [126] supported the idea that populations would always grow until there is a limit of resources. Boserup [29] argued that populations devise new methods for food production when required instead of letting it affect them. This supports the theory that if the economy is doing well the populations would grow and do better. However, this does not seem to happen in real economics, where the broad middle class depicts the wide gap between the upper and lower classes.

TABLE 6.4: Difference between Nash equilibrium and evolutionary stable state.

Nash Equilibrium
(**John Forbes Nash**) Game theory, explains a concept in a game involving more than one player. If players have chosen a certain strategy and no player will gain anything (in its payoff) by changing its strategy, the players have attained Nash equilibrium.
Evolutionary Stable State
(**John Maynard Smith**) Evolutionary game theory, for a given set of behaviors (conserved over time), there must be a profitable action in common, such that no other *mutant* (or new) behavior enters the system at the time. In a game description, if n-players are playing various strategies they adopt one such strategy which is profitable to everyone, such that no other new strategy can be adopted by any player at that time.

6.6   Learning in an Iterated Prisoner's Dilemma Game

The prisoner's dilemma (PD) game is being used in game theory to depict situations of competition or cooperation among players. The game is defined as a non-zero sum game indicating that whenever one player benefits the other player suffers penalties. The players do not have any knowledge of what the other player might play thus this make it a non-cooperative game.

A classical form of the prisoner's dilemma (PD) game is described.

Two suspects are arrested by the police. The police have insufficient evidence for a conviction, and, having separated both prisoners, visit each of them to offer the same deal. If one testifies (defects from the other) for the prosecution against the other and the other remains silent (cooperates with the other), the betrayer goes free and the silent accomplice receives the full 10-year sentence. If both remain silent, both prisoners are sentenced to only six months in jail for a minor charge. If each betrays the other, each receives a five-year sentence. Each prisoner must choose to betray the other or to remain silent. Each one is assured that the other would not know about the betrayal before the end of the investigation. How should the prisoners act? [150]

The game is essentially a two-player game where each player is trying to maximize its own payoff without any consideration of what happens to the other player.

TABLE 6.5: Prisoner sentences in PD game.

	Prisoner B stays silent	Prisoner B betrays
Prisoner A stays silent	Each serves 6 months	Prisoner A: 10 years, Prisoner B: goes free
Prisoner A betrays	Prisoner A: goes free, Prisoner B: 10 years	Each serves 5 years

In a one-shot game, because the players have no knowledge of other player strategies, the game may not be very useful. But in an iterated prisoner dilemma game, the game is repeatedly played amongst players. When repeatedly playing the game, the players have a chance to punish others, if they have played a strategy which was unfavorable to them previously. This is similar to reinforcement learning where, by punishing the player, they can learn the beneficial strategies to play. The game can be repeated infinitely and eventually find an equilibrium, where they learn to play the good defect strategy to prevent being punished in the future. In its classical form, the game presents a Nash equilibrium when the players both defect.

Conducting the game on a trial-by-trial basis or a series of moves, the players must choose either to cooperate or defect on each trial. Table 6.6 shows the numerical payoffs of the strategies played. Table 6.6 depicts a mathematical representation where T stands for temptation to defect, R for reward for mutual cooperation, P for punishment for mutual defection and S for sucker's payoff. In this situation the following inequality will always hold:

$$T > R > P > S \tag{6.10}$$

TABLE 6.6: Payoff matrix in PD game, where R=3, S=0, T=5, P=1.

	Cooperate	Defect
Cooperate	R,R (3,3)	S,T (0,5)
Defect	T,S (5,0)	P,P (1,1)

Playing the game repeatedly will eventually lead to an equilibrium, where all players learn to defect or stay silent to achieve the maximum payoff. This is maintained at a condition where the following rule is true [154, 48]:

$$2R > T + S \tag{6.11}$$

The tendency to defect is the dominating move for the players. But when players jointly defect the payoff returned is less than the payoff returned with mutual cooperation. Playing the game once, clearly the players would think of defecting, but playing it with many trials, the players learn that they have a higher probability of getting a high payoff if they choose to cooperate, eventually trusting the other player to cooperate.

Researchers have used the iterated prisoner's dilemma game to draw important conclusions on behavior of group selection or mutual altruism in real individuals. The gaining of trust among individuals when coming together in groups is often viewed as an evolutionary process which allows evolution of cooperative behaviors. Politics exhibits a PD scenario, illustrating when the country has to make decisions in spending money on its military expansion or reducing weapons. Advertising in economics is viewed as an example of a PD scenario, where firms are competing against each other for sales. They have to decide whether they need to advertise or not depending on whether the other firm has advertised. Their decisions and the times at which they make them would affect their sales.

Miller [132] used automaton to represent a strategy in a prisoner's dilemma game. A player can make only two moves: either to cooperate or defect. A strategy, however, is a complete plan of the number of times to cooperate or defect depending on what the other player played. This can be represented as a sequence of states to determine the next move for each player. For instance, some of the strategies can be as follows:

Always cooperate. Always cooperate no matter what the other player plays (Figure 6.23(a)).

Always defect. Always defect no matter what the other player plays, cooperates or defects (Figure 6.23(b)).

Tit for tat. Cooperate on the first move. Then mimic whatever the other player plays (Figure 6.23(c)).

Figure 6.23 depicts examples of automaton being used to represent the prisoner dilemma strategies. Table 6.7 explains how two players playing an all defecting strategy against a tit-for-tat strategy progress.

The players have no knowledge of what other players might be playing at time $t = 0$. After the players have made their move, they know what the last played strategy was. When an all defecting strategy plays against a tit-for-tat strategy, it starts with the first player playing a defect and the second player cooperating. As a result, the first player benefits getting a better payoff and Player 2 suffers. But after this time step, Player 2 starts to mimic Player 1's last move. Since Player 1 defected in the last time step, it now plays a defect. Player 1 is playing a strategy to defect. Each of these moves returns certain payoffs to the players as shown.

Axelrod [14] organized a prisoner's dilemma tournament where he invited

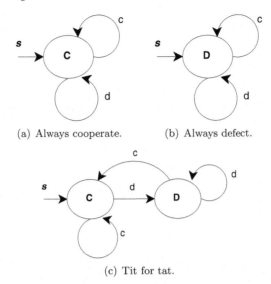

(a) Always cooperate. (b) Always defect.

(c) Tit for tat.

FIGURE 6.23: Example automaton for prisoner's dilemma strategies.

TABLE 6.7: All defect strategy (Player 1) playing against a tit-for-tat strategy (Player 2).

	Player 1 (All D)	**Player 2 (Tit-for-tat)**
At *time* $= t$	D	C
Payoff returned	(5)	(0)
At *time* $= t + 1$	D	D
Payoff returned	(1)	(1)
At *time* $= t + 2$	D	D
Payoff returned	(1)	(1)

game theorists to submit their own strategies for playing the game. Each strategy was played against the other about 200 times and their collected payoffs were collected. The experiment resulted in declaring the 'Tit-for-Tat' [154] strategy as the most successful strategy among the pool of strategies submitted. Jennings et al. [165] introduced an alternate strategy which used a tell to predict the other players' strategy because it is being played a number of times.

In another experiment, Axelrod [12] introduced evolving strategies to play against each other. The results showed that the most effective strategies propagated through the population, initially moving away from cooperation, but then slowly moved towards it again. The average score of the population was also seen to increase as the population evolved to cooperate with each other.

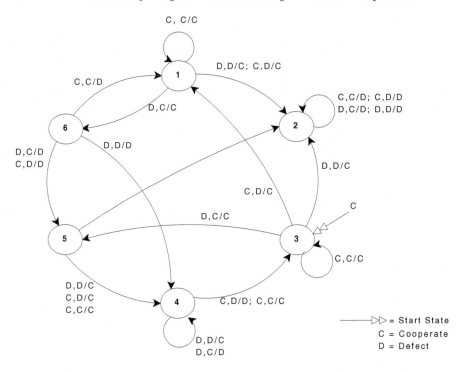

FIGURE 6.24: Finite state machine of eight states representing a prisoner's dilemma strategy. cf. [81].

Fogel [60] implemented a population of coevolving finite state machines (FSM) each with eight states to represent the various strategies of the PD game. Each FSM represented a predictive algorithm for a strategy and were allowed to mutate and evolve in light of the expectation of what the other state machines played. Figure 6.24 shows an example of a Fogel's finite state machine representing a strategy.

In contrast to Axelrod's results of cooperation, Fogel showed that the level of cooperation was not complete in most cases of the machines. His results showed that trials with larger populations, however, did show emergence of cooperative behavior but with smaller numbers and there was "a repeated pattern of initial complete mutual cooperation, but this quickly degenerated into cyclic behavior with moves covering the range from complete cooperation to complete defection" [81]. These experiments were useful to hint the ability of how evolutionary computation can be used to perform problem solving and generate any kind of behavior in simulations [61].

The prisoner's dilemma game allows players to compete against each other to win payoffs. Locations can be used to allow closer players to continuously cooperate or defect to see which strategy wins the most. The players can assess their strategies based on the fitness in the prisoner's dilemma game (Table 6.5).

The strategy played in the prisoner's dilemma game was a 16-state strategy. A 16-state strategy had a similar structure to the design of the automaton discussed by Miller [132].

Table 6.8 represents a three-state automaton represented as a series of strings showing a three state strategy. Figure 6.25 displays the corresponding strategy of this automaton. The starting state is State 0. In this state the player will cooperate. If the other player cooperates the player will move to the State 1, else it will move to State 2. Depending on the new states its next moves will depend on what is represented in the state it is currently in.

TABLE 6.8: Example of a three state machine represented by automaton.

State	C/D	Next State, if Other Player Cooperates	Next State, if Other Player Defects
0	C	1	2
1	D	0	2
2	C	2	2

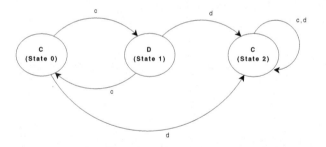

FIGURE 6.25: Example of automaton represented by Table 6.8.

The automaton used in the FLAME iterated prisoner's dilemma game uses a 16-state strategy. A 16-state strategy is represented using 4 bits for each state. Each strategy will contain 16 states; a payoff playing that strategy is the score. Players maintain a database of these strategies in their memory, to aid their competition in the simulation. The structure of the strategy database is pictured in Figure 6.26.

Figure 6.27 shows the structure of one state in this strategy. The state in the strategy is a string of 9 bits. The first bit represents which strategy to play when in this state. In Figure 6.27, the player will cooperate in this state. After doing so, depending on what the other player plays, it will move to a new state. If the other player cooperates, the player will move to the next

FIGURE 6.26: Strategy database of ten strategies in player memory.

FIGURE 6.27: One state in a strategy.

state which is represented by the next 4 bits in the strings or move to the state represented by the last 4 bits, if the other player defected.

Because the length of each strategy was 16 states the game was played 16 times between the players. This ensured that all the states in the strategy were reached during the plays testing the complete strategy.

Using this, ideal payoffs for the players were calculated. If all players started to cooperate, this would be the maximum payoff they will strive to achieve. This will be given as

$$Cooperating\ equilibrium\ =\ Average\ payoff \times 16 = (3+3)/2 \times 16 = 48 \tag{6.12}$$

Similarly the other equilibriums for the other situations will be given as

$$Defecting\ equilibrium\ =\ (1+1)/2 \times 16 = 16 \tag{6.13}$$

$$Mixed\ equilibrium\ =\ (0+5)/2 + (0+5)/2 \times 16 = 2.25 \times 16 = 36 \tag{6.14}$$

Figure 6.28 displays two parents strategies in a three-state automaton. The parent are performing crossover at a point denoted by state number and the length in the state. Therefore as depicted in the Figures 6.28(a) and 6.28(b), the crossover point is at state number = 1 and state length = 4.

Figure 6.29 depicts the two children created by crossing over the two parents. Figure 6.30 depicts a mutant child of Parent 1 which was mutated at the same position. The diagrams show how through crossover and mutation techniques, new strategies can be generated by just moving the bits in the strings. Table 6.9 summarizes the values used during the experiment.

Steps taken in the PD model:

- Step 1: Citizen agent chooses a chosen strategy it might play using the roulette wheel selection mechanism. Posts the first step which is either to cooperate (C) or defect (D).

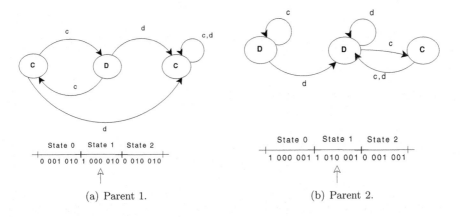

(a) Parent 1.   (b) Parent 2.

FIGURE 6.28: Two strategies acting as parents.

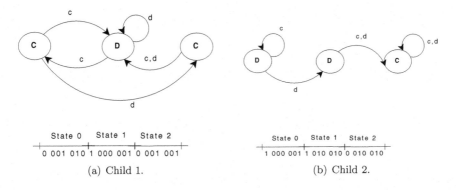

(a) Child 1.   (b) Child 2.

FIGURE 6.29: Two children resulting from crossover of parents, at crossover point state number 1 and state length 4.

- Step 2: Citizen agent performs crossover and mutation techniques on the strategy for the PD game.

- Step 3: Solver agent reads in the strategies of the two players and plays the game between them. Adds the payoffs collected and tells the citizen about the outcome, who won and who lost.

Figure 6.31 depicts the average score when the payoff of the IPD game is used as the score of the strategy. The graphs were plotted with their ideal values, in Equations 6.12 - 6.14. This shows which equilibrium was favorable for the players. In Figure 6.31, the players were seen to learn the equilibrium values very quickly in the simulation. The payoffs varied between 40 and 80, but stabilized above the ideal cooperating equilibrium.

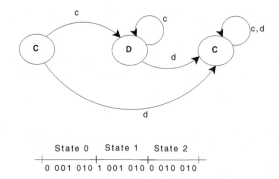

FIGURE 6.30: Mutant child of Parent-1 at mutation point state number 1 and state length 4.

TABLE 6.9: Numerical values in FLAME-IPD experiment.

Variable	Value
Landscape	200 x 200
n	50 (Number of players)
Coop*	48 (Cooperating equilibrium)
Def*	16 (Defecting equilibrium)
Mixed*	36 (Mixed equilibrium)

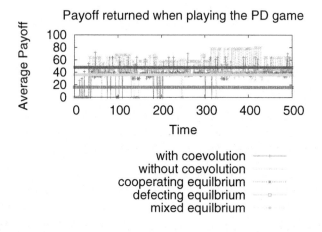

FIGURE 6.31: Score is payoff returned playing IPD game.

6.7 Multi-Agent Systems and Games

Modeling complex system behavior is an emergent science which demonstrates the complex social behavior of different communities working together. Multi-agent systems can be used effectively to model this. These systems are inherently distributed where agents are either spatially spread heterogeneous in nature and have limited information available to them. Multi-agent systems are essentially players involved in a non-cooperative game scenario. If all the individuals tried to optimize their behavior, globally the system may optimize as well. But there are problems analyzing these optimizations.

1. It is not possible to have a payoff matrix for models which are not games to begin with. The measurement of the payoff will have to be associated with a fitness variable U as part of the agent memory.

2. Evolutionary algorithms are used within agents, primarily to allow them to evolve. Most of these algorithms use a comparison technique to calculate the difference between the actual fitness collected and the expected fitness of a particular strategy. In multi-agent systems, the expected fitness is difficult to predict until the agent has tried the strategy in the simulation.

3. Multi-agent systems are sometimes deterministic or stochastic. Agents can use their memories to save good strategies, making the model deterministic by knowing what to play next. However, there is unpredictable behavior in complex systems and agents struggle to find the better strategies. The system then takes longer to reach global maxima.

4. Agents need to calculate when to change their behavior to reduce complexities in their code.

5. The strategies are sometimes a continuous variable and not a set of discrete strategies as in traditional game theory.

Importing the principles from game theory into multi-agent systems would thus require a number of assumptions to be embedded into the agents to test the theory. The world economy is often referred to as an "evolving game with nearly four billion players" [49].

Multi-agent systems have a very close relation to the principles of the games. The works of [13] and [73] are a few examples where game theory has been used to develop agent-based models of players playing games and evolving characteristics [132].

The behavior of an agent is programmed by the way it should respond to signals using rules embedded in the program. Signals are messages coming from other agents or can be the changes in the environment which affect the agent's behavior. The agent's behavior is termed as the *strategies* or functions the agent can perform. These strategies are specific to certain decisions or general. Programmatically these strategies are represented as rules and are converted into functions to calculate the next move of the agent. There are two research streams intp which similar agent games can be divided:

Competitive Equilibrium Theory. Agents only respond to signals from the environments, like how the price affects the agent's buying capability.

Computational Mechanism Design. Assumes agents respond in a game theoretic manner by modeling the effect of their actions on actions of other agents in the system.

In multi-agent systems, the two research streams have to be combined into one, because agents are not only responding to environment conditions but other agents in the system as well.

Chapter 7

Agents in Biology

7.1   Example Models 176
    7.1.1   Molecular Systems Models 176
    7.1.2   Tissue and Organ Models 179
    7.1.3   Ecological Models 182
    7.1.4   Industrial Applications of Agent-Based Modeling with
         FLAME 183
7.2   Modeling Epithelial Tissue 184
    7.2.1   Merging with Other Toolkits 185
7.3   Modeling Drosophila Embryo Development 187
    7.3.1   Stochastic Modeling 188
    7.3.2   Converting to an Agent-Based Model 188
    7.3.3   Find Optimum Model Settings 196
7.4   Output Files for Analysis 198
7.5   Modeling Pharaoh's Ants (*Monomorium pharaonis*) 202
7.6   Model Drug Delivery for Cancer Treatment 224
    7.6.1   Using Multiple Outputs 234

Biological systems are often a collection of multiple complex systems. These systems range from small bacterial models or large cell tissue models and their behavior with other organs. Complex systems display adaptive behavior to continually changing environment, coping to survive. These systems are extremely robust. Studying these systems is extremely cumbersome, due to their complexity, size and capability of collecting data at minute scales. Simulation, however, allows biologists to conceptually visualize how these systems function and what factors affect them. Having described these system as a series of steps in models, the biologists can then test the data produced through simulation with real data, inherently matching their predictions and understanding to the real systems.

Describing a biological system virtually thus involves the following:

- Make design decisions: Identifying the system functions of the model being simulated. Following agile methods, this process involved repeated conversations between domain experts (i.e. biologists) and computer scientists (i.e. programmers).

- List agents and functions: Identify agent states and the order in which they function during one iteration in the simulation.

- Function programming: Develop the agent function code.

- Messages between agents: Through the design phase, the input and output messages involved with the agents need to be identified and then linked with the functions.

- Determine agent memory: For each agent, identify the memory variables. These will form part of the function code, being manipulated through messages and agent functions during the simulation.

- What to measure: Identify the model output variables that will be recorded as simulation objectives. These output variables can either be average agent behavior on one variable or a number of variables which change during the simulation to understand agent behavior against iteration time. These outputs can then be compared to real system data to compare how accurate the model represents the real system.

FLAME has been very successful in modeling a variety of biological experiments. Working with various biologists and involved in their projects, it has studied systems such as epithelial tissue healing, bacterial concentrations in oxygen-starved environments, ant and pheromone behavior and even sperm behavior in reproductive systems. It has aided in unlocking interesting biological phenomena, just by the exercise of conceptualizing and writing models, with comparing simulated to real collected data. Some of these projects, in collaboration with experimental biologists, are summarized below.

7.1   Example Models

7.1.1   Molecular Systems Models

Innate immune system. NFκB pathways and its relationship with the cytoskeleton. Nature is governed by local interactions among lower-level subunits, whether at the cell, organ, organism or colony level. Adaptive system behavior emerges via these interactions, which integrate the activity of the subunits. To understand the system level it is necessary to understand the underlying local interactions. Successful models of local interactions at different levels of biological organisation, including epithelial tissue and ant colonies, have demonstrated the benefits of such 'agent-based' modeling. Here, the modelers presented an agent-based approach to modeling a crucial biological system, the intracellular NFκB signalling pathway. The pathway is vital to immune response regulation, and is fundamental to basic survival in a range of species. Alterations in pathway regulation underlie a variety of diseases, including atherosclerosis and arthritis. The modeling of individual molecules, receptors and genes provides a more comprehensive outline of regulatory network mechanisms than previously possible with equation-based approaches. The method also permits consideration of structural parameters in pathway regulation. The modelers predicted that inhibition of NFκB is directly affected by actin filaments of the cytoskeleton sequestering excess inhibitors, therefore regulating steady-state and feedback behavior [149].

Computational modeling of NFκB activation using IL-1RI and its co-receptor TILRR, predicts a role for cytoskeletal sequestration of IκBα in inflammatory signalling. The transcription factor NFκB is activated by toll-like receptors and controlled by mechanotransduction and changes in the cytoskeleton. In this study we combine 3-D predictive protein modeling and in vitro experiments with in silico simulations to determine the role of the cytoskeleton in regulation of NFκB. Simulations used a comprehensive agent-based model of the NFκB pathway, which includes the type 1 IL-1 receptor (IL-1R1) complex and signalling intermediates, as well as cytoskeletal components. Agent-based modeling relies on in silico reproductions of systems through the interactions of its components, and provides a reliable tool in investigations of biological processes, which require spatial considerations and involve complex formation and translocation of regulatory components. The modelers showed that their model faithfully reproduced the multiple steps comprising the NFκB pathway, and provided a framework from which they can explore novel aspects of the system. The initial analysis, using 3D predictive protein modeling and in vitro assays, demonstrated that the inhibitor IκBα is sequestered to the actin/spectrin complex within the cytoskeleton of the resting cell, and released during IL-1 stimulation, through a process controlled by the IL-1RI co-receptor TILRR. In silico simulations using the agent-based model predict that the cytoskeletal pool of IκBα is released to adjust signal amplification in relation to input levels. The results suggest that the process provides a mechanism for signal calibration and enables efficient, activation-sensitive regulation of NFκB and inflammatory responses [161].

MapKinase pathways. Signal transduction through the Mitogen Activated Protein Kinase (MAPK) pathways. Many cells use these pathways to interpret changes to their environment and respond accordingly. The pathways are central to triggering diverse cellular responses such as survival, apoptosis, differentiation and proliferation. Though the interactions between the different MAPK pathways are complex, nevertheless, they are capable of maintaining a high level of fidelity and specificity to the original signal. There are numerous theories explaining how fidelity and specificity arise within this complex context; spatiotemporal regulation of the pathways and feedback loops are thought to be very important. This experiment presents an agent-based computational model addressing multi-compartmentalization and how this influences the dynamics of MAPK cascade activation. The model shows that multi-compartmentalization coupled with periodic MAPK kinase (MAPKK) activation may be critical factors for the emergence of oscillation and ultrasensitivity in the system. The model establishes a link between the spatial arrangements of the cascade components and temporal activation mechanisms, and how both contribute to fidelity and specificity of MAPK-mediated signalling.

Oxidase regulation in anaerobic *E. Coli* cells. In the presence of oxygen (O2) the model bacterium *Escherichia coli* can conserve energy by aerobic respiration. Two major terminal oxidases are involved in this process, (1) Cyo has a relatively low affinity for O2 but is able to pump protons and hence is energetically efficient, and (2) Cyd has a high affinity for O2 but does not pump protons.

When *E. coli* encounters environments with different O2 availabilities, the expression of the genes encoding the alternative terminal oxidases, the cydAB and cyoABCDE operons, are regulated by two O2-responsive transcription factors, ArcA (an indirect O2 sensor) and FNR (a direct O2 sensor). It has been suggested that O2-consumption by the terminal oxidases located at the cytoplasmic membrane significantly affects the activities of ArcA and FNR in the bacterial nucleoid. In this study, the agent-based modeling approach represented spatially the bacterial process and simulated the uptake and consumption of O2 by *E. coli*. It also presented a consequent modulation of ArcA and FNR activities based on experimental data obtained from highly controlled chemostat cultures. The molecules of O2, transcription factors and terminal oxidases were treated as individual agents and their behaviors with interactions were imitated in a simulated 3D *E. coli* cell. The model implied that there are two barriers that dampen the response of FNR to O2, i.e. consumption of O2 at the membrane by the terminal oxidases and reaction of O2 with cytoplasmic FNR. Analysis of FNR variants suggested that the monomer-dimer transition is the key step in FNR-mediated repression of gene expression [17].

Blood-brain barrier and nanoparticles. Blood mediated nanoparticle delivery is a new and growing field in the development of therapeutics and diagnostics. Nanoparticle properties such as size, shape and surface chemistry can be controlled to improve their performance in biological systems. This enables modulation of immune system interactions, blood clearance profile and interaction with target cells, thereby aiding effective delivery of cargo within cells or tissues. Their ability to target and enter tissues from the blood is highly dependent on their behavior under blood flow. Here, the modelers produced an agent-based model of nanoparticle behavior under blood flow in capillaries. They demonstrated that red blood cells are highly important for effective nanoparticle distribution within capillaries. Furthermore, this model demonstrated how nanoparticle size can selectively target tumor tissue over normal tissue. The authors showed that the polydispersity of nanoparticle populations is an important consideration in achieving optimal specificity and to avoid off-target effects. In the future, this model could be used for informing new nanoparticle design and to predict general and specific uptake properties under blood flow [71].

7.1.2 Tissue and Organ Models

Epithelial tissue and wound healing. Transforming growth factor (TGF-$\beta 1$) is a member of the TGF-beta superfamily ligand-receptor network. It plays a crucial role in tissue regeneration. The extensive in vitro and in vivo experimental literature describing its actions nevertheless describe an apparent paradox in that during re-epithelialization it acts as proliferation inhibitor for keratinocytes. The majority of biological models focus on certain aspects of TGF-$\beta 1$ behavior and no one model provides a comprehensive story of this regulatory factor's action. Accordingly the model's aim was to develop a computational model to act as a complementary approach to improve our understanding of TGF-$\beta 1$. In a previous study, an agent-based model of keratinocyte colony formation in 2D culture was developed. In this study, the model was extensively developed into a 3D multiscale model of the human epidermis which is comprised of three interacting and integrated layers:

- An agent-based model which captures the biological rules governing the cells in the human epidermis at the cellular level and includes the rules for injury-induced emergent behaviors.

- A COmplex PAthway SImulator (COPASI) model which simulates the expression and signalling of TGF-$\beta 1$ at the sub-cellular level.

- A mechanical layer embodied by a numerical physical solver responsible for resolving the forces exerted between cells at the multicellular level.

The integrated model was initially validated by using it to grow a piece of virtual epidermis in 3D and comparing the in virtuo simulations of keratinocyte behavior and of TGF-$\beta 1$ signalling, with extensive research describing the key regulatory protein. This research reinforced the idea that computational modeling can be an effective additional tool to aid understanding of complex systems. In the paper produced, the model was used to explore the hypotheses on functions of TGF-$\beta 1$ at the cellular and subcellular level on different keratinocyte populations during epidermal wound healing [2].

Oviduct and sperm motility. The processes by which individual sperm cells navigate the length and complexity of the female reproductive tract, to reach and fertilize the oocyte, are extremely fascinating and difficult to study. Numerous complex processes can potentially influence the movement of spermatozoa within the tract, resulting in a regulated supply of spermatozoa to the oocytes at the site of fertilization. Despite significant differences between species, breeds and individuals, these processes converge, ensuring that an optimal number of high quality spermatozoa reach the oocytes, resulting in successful fertilization without a significant risk of polyspermy. Computational modeling provides a useful method to combine knowledge about the individual processes to help understand the relative significance of each factor. In this study, the first agent-based computational model of sperm behavior within an oviductal environment was created. Firstly, a generic conceptual model of sperm behavior within the 3D oviduct was presented. Sperms are modeled as individual cells with a set of behavioral rules defining how they interact with their local environment and regulate their internal state. Secondly, a set of 3D models of the mammalian oviduct were constructed. Histology images of the mouse oviduct were obtained and the path that the oviductal tube follows through the tissue was identified using CUDA-based image analysis (using GPUs). This was used to determine cross-sectional topology, and measurements from the cross sections were used to generate a set of accurately scaled 3D models of the oviduct. The process of constructing and validating the agent-based computational model of sperm movement and transport within the oviductal environment was described. The model is grounded in reality, with accurate space and time scales used throughout, and parameters and mechanisms from literature where available. Sensitivity analysis was performed on all parameters, and those most sensitive to variation were identified. The model was validated against literature, to validate it. However the model had a few limitations based on the assumptions drawn which were also presented. The model was used to investigate the significance of the oviductal environment on the regulation of sperm distribution and their progression to the site of fertilization. How changes to the the oviduct environment can alter the sperm distribution was also studied. Finally, the potential use for the model and how more complex mechanisms could be integrated in the future were discussed [34].

Blood flow. This work investigated how a specific biological system - heart cells and tissue - can be studied using computation as a metaphor, and addressed the question of whether biological behavior can be labelled as a computation. Clearly, an answer to this question is an ambitious goal and is yet far off; however, the intent was to take a small step towards it. As a test-bed, this work aimed to describe and implement a novel computational perspective for modeling cardiac electrophysiology. It aimed specifically to develop a hybrid, hierarchical, agent-based model of the cardiac cell and tissue electrophysiology. The model draws upon and extends the formal computational paradigms of hierarchy, state machines and hybrid models to simplify model development; but more importantly, to accurately simulate, verify and validate the system against the more traditional models that use numerical methods [197].

Epithelial renewal and long-term survival. Epithelial renewal in skin is achieved by a constant turnover and differentiation of keratinocytes. Three popular hypotheses were proposed to explain basal keratinocyte regeneration and epidermal homeostasis:

- asymmetric division (stem-transit amplifying cell);
- populational asymmetry (progenitor cell with stochastic fate); and
- populational asymmetry with stem cells. In this study, the lineage dynamics was investigated using these hypotheses with a 3D agent-based model of the epidermis.

The model simulated the growth and maintenance of the epidermis over three years. The offspring of each proliferative cell was traced. While all lineages were preserved in asymmetric division, the vast majority was lost when assuming populational asymmetry. The third hypothesis provided the most reliable mechanism for self-renewal by preserving genetic heterogeneity in quiescent stem cells, and also inherent mechanisms for skin ageing and the accumulation of genetic mutation [117].

Cell and chemical interactions in 3D using HPC for chemotaxis. The behavior of biological cells within the body is far from static; they interact with their environment and each other using chemical secretions which act as signals. Existing tools allow for complex behavior of cells to be modeled, but do not provide built-in mechanisms for handling the chemical communication that occurs. Here, a set of extensions were made to the FLAME agent-based modeling framework to perform 3D chemical diffusion within a constrained environment, and allow individually modeled biological cells to interact with the 3D chemical field by secreting, detecting and consuming different chemicals. FLAME, which automatically parallelized agent-based models, was extended to allow chemical diffusion and automatically performed using an attached GPU. The framework was enhanced to allow the chemical diffusion to be performed not only on a local computer, but also on the GPU nodes of a high performance cluster, while the agents themselves were processed on normal CPU nodes. To validate the technique, two different studies were performed, one looking at the survival of eosinophil cells in the presence of (IL5), and the other looking at eosinophil chemotaxis. Both studies were validated against published experiments.

Modeling the effect of CRTH2 receptor blocker on eosinophilic inflammation during an asthma attack. Eosinophillic inflammation in the lungs is a key symptom of asthma, which occurs in the late phase of an asthma attack, resulting in more severe airway restriction and long-lasting secondary effects. This work presented a model of eosinophilic inflammation, that occurs in lungs during an asthma attack, and the effect that a proposed treatment (i.e. blocking the CRTH2 receptors on resident T-cells) can have on the inflammation. The model starts with the mast cell degranulation associated with initiation of an asthma attack, and the subsequent release of chemicals (PGD2, IL5) from mast cells. The PGD2 stimulates tissue resident T-Cells via the CRTH2 receptor to produce a large quantity of IL5, which has been linked to eosinophil survival. After a few hours, eosinophils, which are produced in large quantities in the bone marrow, flood into the tissue and are chemotactically attracted to the lumen by Eotaxin, released from epithelial cells. When the CRTH2 receptor is blocked, the resident T-cells do not produce additional IL5, therefore reducing the levels of IL5 within the tissue. This influences the amount of time eosinophils are able to survive within the tissue, thereby reducing the severity of an eosinophilic inflammation. The results are validated against experimental data for a related anti-IL5 drug.

7.1.3   Ecological Models

Social insects: foraging in ants. Communication improves decision- making for group-living animals, especially during foraging, facilitating exploitation of resources. Here a model was created of the trail-based foraging strategy of Pharaoh's ants to understand the limits and constraints of a specific group foraging strategy. To minimize assumptions, the model parameters acquired through behavioral study were used. Pharaoh's ants (*Monomorium pharaonis*) exploit the geometry of trail network bifurcations to make U-turns if they are walking the wrong way. However, 7% of foragers perform apparently incorrect U-turns. These seemingly maladaptive U-turns are performed by a consistent minority of specialist U-turners that make frequent U-turns on trails and lay trail pheromones much more frequently compared to the rest of the colony. The study showed a key role for U-turning ants in maintaining the connectivity of pheromone trails. The authors produced an agent-based model of a heterogeneous ant community where 7% of agents were specialized frequent U-turners whilst the remaining 93% rarely U-turned. Simulations showed that heterogeneous colonies enjoyed significantly greater success at foraging for distant food resources compared to behaviorally homogeneous colonies. The presence of a cohort of specialized trail-layers maintained a well-connected network of trails that ensured that food discoveries are rapidly linked back to the nest. This decentralized information transfer might ensure that foragers can respond to dynamic changes in food distribution, thereby allowing more individuals in a group to benefit by successfully locating food finds [164, 97, 96].

Social insects: soil disposal organization. Colonies of *Pheidole ambigua* ants excavate soil and drop it outside the nest entrance. The deposition of thousands of loads leads to the formation of regular ring-shaped piles. But, *how is this pattern generated?*

This study investigated soil pile formation on level and sloping surfaces, both empirically and using an agent-based model. The authors found that ants drop soil preferentially in the direction in which the slope is least steeply uphill from the nest entrance, both when adding to an existing pile and when starting a new pile. Ants respond to cues from local slopes to choose downhill directions. Ants walking on a slope increase the frequency and magnitude of changes in direction, and more of these changes of direction take them downhill than uphill. Also, ants carrying soil on a slope wait longer before dropping their soil compared to ants on a level plane. These mechanisms combine to focus soil dropping in the downhill direction, without the necessity of a direct relationship between slope and probability of dropping soil. These empirically determined rules were used to simulate soil disposal. The slight preference for turning downhill measured empirically was shown in the model to be sufficient to generate biologically realistic patterns of soil dumping when combined with memory of the direction of previous trips. From simple rules governing individual behavior, an overall pattern emerges, which is appropriate to the environment and allows a rapid response to changes [163].

Further general titles and experiments can be found in [86, 87].

7.1.4 Industrial Applications of Agent-Based Modeling with FLAME

Active management of crowds in airports, stations and shopping malls. This used the Concoursia software application, based on FLAME-GPU. This models individual people moving around precisely modeled buildings, used for both planning buildings and also for actively managing crowds when connected to suitable sensor systems. This provides managers with a decision support tool for making decisions about potential interventions to deal with overcrowding based on predictions provided by the FLAME model (Figure 7.1(a)).

Managing patient demand in hospital A&E departments. This uses the application PatientFlow, based on FLAME running on an HPC grid. In the model, each patient, staff member and ancillary service is modeled as individuals, to present a detailed representation of how the hospital

 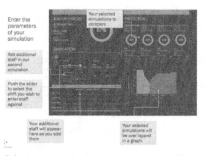

(a) A view of the Concoursia simulation of a main line railway station.

(b) A screenshot from the PatientFlow app.

FIGURE 7.1: Industrial applications of FLAME.

operates. The initial model was developed in collaboration with Central Manchester Foundation Trust and the Science and Technology Facilities Council. The system can make predictions of where bottlenecks are likely to occur in the hospital over the next time period of up to 48 hours and can provide clinicians and managers with advice about how to manage the demands in order to reduce waiting times and more (Figure 7.1(b)).

7.2   Modeling Epithelial Tissue

Normal human keratinocyte (NHK) cells form over 80% of the outermost layer of the skin or the epidermis. The epidermis is a fast renewing tissue which forms a protective barrier between our internal organs and the outside world. Understanding how cells proliferate and self-organize into layers of skin tissue is a very important research topic. Such understanding promotes the development of methods to artificially produce reconstructed human skin for patients with heavy skin loss, such as chronic burns, wounds or skin disease. As part of the Epitheliome project, Sun and McMinn used FLAME to develop an in-virtuo model of skin cell behavior. The interaction of the software agents in the in-virtuo model described the NHK macroscopic morphogenesis in-vitro [192]. Figure 7.2 shows a comparison of pictures from real and simulated images.

In the model of the keratinocyte colony formation, each cell was represented as an individual agent. The signaling process between the cells was simulated by reading and writing messages between agents. The model algorithm for the keratinocyte colony formation is given as follows. These functions were performed in one time step but represented 30 minutes of real time (Figure 7.3).

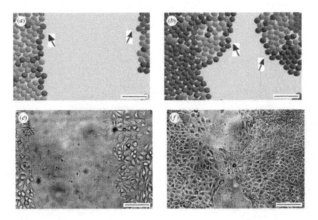

FIGURE 7.2: Comparing real and simulated data of wound healing in 2D. Adapted from [192].

1. Cells communicate with each other by exchanging information about their types and locations. In the X-agent representation, the cell agent sends a signal message such as cell location.

2. Cells act accordingly and go into or continue a cell cycle which includes several checkpoints. In the X-agent, the cell performs a function consisting of cell cycles and rules.

3. Cells divide depending on certain conditions (location and calcium concentration in the environment). The cell X-agent divides into two by producing another cell with a unique identifier.

4. Cells differentiate depending on certain conditions (type, location and calcium concentration in the environment). The cell X-agent performs differentiation functions.

5. Cells migrate depending on certain conditions. The cell X-agent moves to a close-by location depending on its calculations.

6. Add a physical solver to prevent cells from overlapping or occupying the same space.

7.2.1 Merging with Other Toolkits

The FLAME kerotinocyte model involved each agent to read positions of neighboring cells and perform internal functions based on these. These activities such as differentiation or complex calculations can sometimes be done in

(a) At time step 0 (Beginning).

(b) At time step 30.

(c) At time step 60.

(d) At time step 90.

(e) At time step 130.

FIGURE 7.3: 3D model snapshots of wound healing at different time steps of the simulation.

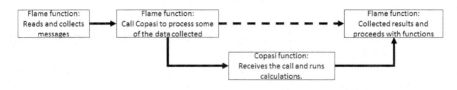

FIGURE 7.4: Calling Copasi from FLAME C code.

other toolkits and languages better than writing them as C functions. For this purpose, it makes sense for the simulation code in C to halt and call other toolkits, receive the results and then progress with the results. In this case COPASI was called through C commands to complete part of the simulation work (Figure 7.4). This method can also be merged with MATLAB and other toolkits when doing complex matrix calculations which are cumbersome to do in the C language.

7.3 Modeling Drosophila Embryo Development

In this experiment, two models of the same phenomenon were compared - one written using stochastic methodology in MATLAB and other using agent-based modeling. The Bicoid morphogen gradient establishment takes place during early embryo development in *Drosophila melanogaster*, and is a dynamic system that allows the Bicoid molecules to diffuse along the embryo anterior-posterior (A-P) axis in different developmental stages. The protein concentration gradient is sensed by downstream genes and induces differential spatial pattern of gene expression. In most analyses of this process, the bicoid mRNA is thought to supply proteins at a constant rate in the anterior pole of embryo. Based on these experimental evidences, Liu and Niranjan [120] proposed three Bicoid concentration computational models in which the maternal bicoid mRNA is regulated by being held constant for 2 hours and followed by rapid decay. The uncertainty of such source regulation model is also verified later by Gaussian processing in [121].

In this work, two approaches of modeling Bicoid morphogen concentration gradient are compared. The first approach, stochastic chemical reaction system [120, 121], models the propagation of the diffusion rate of changes using the stochastic modeling in MATLAB software. Another model in agent-based modeling using FLAME is implemented to compare the results and the problems faced in both modeling approaches.

7.3.1   Stochastic Modeling

The stochastic Bicoid protein reaction diffusion system was implemented as a simulation of 100 compartments along the A-P axis, each with length $h = 5\mu m$, which is the average size of one nucleus. The proteins diffuse along the compartments based on time and concentrations.

$$Bicoid_1 \leftrightharpoons Bicoid_i \leftrightharpoons Bicoid_n \qquad (7.1)$$

where $\leftrightharpoons$ represents the diffusion for $i = 1$ to n. The Bicoid protein has a life and degrades in all compartments along the axis, until it translates into bicoid mRNA in compartment 1 to form the anterior pole of embryo. Details of the procedure are as follows:

Algorithm: Bicoid reaction-diffusion stochastic simulation

Input: Model parameters; final time.

Output: Bicoid molecular numbers along 100 compartments: m.

Start $m = 0$; $t = 0$;

Repeat:

1. Generate two random numbers which are uniformly distributed in $(0, 1)$: $r(1)$ and $r(2)$.
2. Calculate propensity functions of all the reactions: $a = a1 + a2 + a3 + a4$.
3. Calculate the time when next reaction occurs: $t + \tau$, where $\tau = 1/a \; ln(1/r(1))$.
4. Decide which reaction occurs at $Pt + \tau$: find $j \epsilon R$ such that $\Sigma_{i=1}^{j-1} a_i/a \le r(2) < \sum_{i=1}^{j} a_i/a,$
5. Update numbers of reactants and products in j-th reaction and set $t \leftarrow t + \tau$ until $time > finaltime$.

The diffusion between neighboring compartments takes place in both directions based on a rate d, related to a diffusion constant of a deterministic model $d = D/h2$. The vector m contains a number of molecules along the $N = 100$ compartments or bins. This is based on the equations in [121].

7.3.2   Converting to an Agent-Based Model

Figure 7.5 shows a structured view of the embryo to understand how a protein diffuses through the length of the embryo. As assumed with the stochastic model, the embryo cell is divided into 100 compartments, with the source in the first compartment. The source produces proteins at a certain rate r, and depending on a rate d, individual proteins diffuse into the next compartment to move forward down the length of the embryo, depending on the concentration gradients across the membranes. Following are the steps to the agent model.

Algorithm 2: Bicoid reaction-diffusion agent-based simulation

Input: Model parameters (agent memory variable values) at time $t = 0$; final time of simulation is equal to number of iterations.

Output: Bicoid molecular numbers along 100 compartments: m.

Start m $= 0$; t $= 0$;

Repeat:

1. Generate protein production rate as uniformly distributed in $(0, 1)$.
2. If source not decayed, calculate probability of producing the protein.
3. Decide which reaction occurs allowing molecules to move to the left or right.
4. Update numbers of molecules in each compartment.
5. Until $time > number of iterations$.

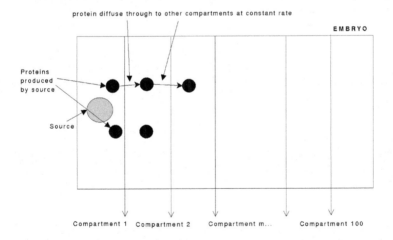

FIGURE 7.5: Movement of proteins within a Drosophila embryo. A structured view.

In both modeling techniques, it is best to start with the problem, decompose it to simpler sub-problems, and then solve each sub-problem separately. Therefore rather than using the stochastic model as a starting point, modeling is easier if we start with the scenario being modeled and then representing this as agents to compare results later.

- Source agent. The source agent is producing new protein agents. The source will also be decaying after a certain time period to reduce its life and eventually disappear or stop producing proteins.

- Protein agent. The protein agents are diffusing across the compartment depending on the protein concentration in the compartment it is in.

- Compartment agent. To depict the embryo as a whole, a hundred compartment agents can be used to determine the concentration of proteins within them. This agent can be avoided as the compartments themselves are not doing any function, but can be used to account for the result analysis later. Alternatively, one agent representing the environment can also be used, thus showing that it depends on modeler perspectives on how the model is written.

Figure 7.6 shows the activities during one iteration of the model. In addition, messages such as the Protein posts location, which are outputs a protein location message which is read by other proteins or compartment agents, not shown. This message would be the input to the functions for Protein agent to calculate the next move or for the Compartment agent to count how many proteins it has.

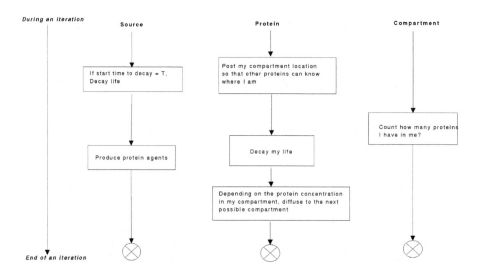

FIGURE 7.6: Agent activities during one iteration.

The model was executed for 12,000 iteration steps, assuming one time step represents one second of the diffusion model (in the stochastic method), with

12,000 iteration steps representing two hours of the stochastic equation model. The other rates used in the model were as follows:

- SOURCE DECAY RATE 0.01 - The decay rate of the source.

- SOURCE TIME DECAY 8640 - The time step at which source will start decaying represents 144 minutes of the stochastic simulation.

- SOURCE TIME PRODUCE 50 - The time after which the source will start producing a protein.

- SOURCE PRODUCTION PROB 1.0 - The production probability of a protein

- PROTEIN DECAY RATE 0.01 - The decay rate of the protein.

- COMPARTMENT DIMENSION X 5 - The dimension width of a compartment.

- COMPARTMENT DIMENSION Y 15 - The dimension height of a compartment.

- PROB RIGHT 0.5 - The probability of a protein to move right to the compartment on the right.

- PROB LEFT 0.1 - The probability of a protein to move left to the compartment of the left. The probability to move right was kept higher as this would be more favourable.

- DIE 0.001 - Numerical value to denote when a life goes below this value, kill the agent.

The values for probabilities of source production rates, protein decay rates, right and left movement probability are conditions that will be changed which each experimental run. Changing these conditions can allow results to be mapped to the stochastic results, in order to find optimum rates which produce the same results. The code for the agent functions is as follows:

```
/**** Compartment Agent Functions **********/
int Compartment_calculateproteins()
{
  int protein_ct=0;
  protein_location_message=get_first_protein_location_message();
  while(protein_location_message)
  {
    if((protein_location_message->y>Y0)&&(protein_location_message->y<Y1))
    {
      protein_ct++;
    }
    if(protein_location_message->compartment==ID)
{
  protein_ct++;
```

```
}
protein_location_message=get_next_protein_location_message
                                    (protein_location_message);
  }

  NUMBEROFPROTEINS=protein_ct;
  printf("\nMY proteins %d , %d", ID, NUMBEROFPROTEINS);
  add_proteinnumber_compartment_message(ID,NUMBEROFPROTEINS);
  return 0;
}

/**** Protein Agent Functions***/
int Protein_post_details()
{
  add_protein_location_message(ID, X,Y, COMPARTMENT_ID);
  return 0;
}

int Protein_decay()//reduce life of protein
{
  if(LIFE<=DIE)
  {
    printf("KILL PROTEIN******");
    return 1;
  }
  return 0;
}

int Protein_calculate_movement()
{
  int next_id,before_id;
  double temp_prob=0.0;
  int temp_time=0;
  int now=0;
  int protein_next,protein_before;
  int flag_before=0;
  int flag_next=0;
  int my_compartment=0;

  temp_prob=random_double(0.0,1.0);
  temp_time=random_int(1,10);

  now=temp_time+TIME_LAST_MOVE;
  protein_before=0;
  protein_next=0;

  flag_before=0;
  flag_next=0;

  if(COMPARTMENT_ID==1)
  {
    flag_before=1;
  }
  if(COMPARTMENT_ID==100)
  {
    flag_next=1;
  }
```

```
my_compartment=0;
proteinnumber_compartment_message=
                    get_first_proteinnumber_compartment_message();
while(proteinnumber_compartment_message)
{
  if(proteinnumber_compartment_message->compartment_id==COMPARTMENT_ID)
  {
    my_compartment=proteinnumber_compartment_message->production;
  }
  if(flag_before==0)
  {
    before_id=COMPARTMENT_ID-1;
    if(proteinnumber_compartment_message->compartment_id==before_id)
    {
      protein_before=proteinnumber_compartment_message->production;
    }
  }

  if(flag_next==0)
  {
    next_id=COMPARTMENT_ID+1;
    if(proteinnumber_compartment_message->compartment_id==next_id)
    {
      protein_next=proteinnumber_compartment_message->production;
    }
  }
  proteinnumber_compartment_message=
                    get_next_proteinnumber_compartment_message
                            (proteinnumber_compartment_message);
}

if(now<=TIME_COUNTER)
{
  if(temp_prob<PROB_LEFT)//check to move left or right
  {
    if(flag_before==0)
    {
  if(my_compartment<protein_before)
      {
        X=X-5;
        Y=Y-random_int(1,10);
        if(Y>15)
        {
          Y=Y-5;
        }
        COMPARTMENT_ID=before_id; // moving left
      }
    }
  }
  else if(temp_prob<PROB_RIGHT)
  {
    if(flag_next==0)
    {
    if(protein_next<my_compartment)
      {
        X=X+5;
```

```
           Y=Y+random_int(1,10);
           if(Y>15)
           {
             Y=Y-5;
           }
           COMPARTMENT_ID=next_id;//moving right
         }
       }
     }
     TIME_LAST_MOVE=TIME_COUNTER;
  }
  TIME_COUNTER++;
  return 0;
}

/**** Source Agent Functions***/
int Source_basic_posts()
{
  TIMECOUNTER++;
  return 0;
}

int Source_idle()
{
  return 0;
}

int Source_decaying()
{
  double temp_double=0.0;
  if(TIMECOUNTER>SOURCETIMEDECAY)
  {
    temp_double=LIFE*SOURCEDECAYRATE;
    LIFE=LIFE - temp_double;
  }
  return 0;
}

int Source_producing()
{
  double temp=0.0;
  double source_production=0.0;
  double temp_time=0.0;
  int i=0;
  if(TIMECOUNTER>SOURCETIMEPRODUCE)
  {
    temp=random_double(0.0,1.0);
    if(TIMECOUNTER>SOURCETIMEDECAY) //OLD case
    {
      temp_time=TIMECOUNTER-SOURCETIMEDECAY;
      temp_time=-1*(temp_time/540);
      source_production=CONSTANT * exp(temp_time);
    }
    else
    {
      source_production=SOURCE_PRODUCTION_PROB;
    }
```

```
      if(temp<=source_production)
      {
        if(LIFE>DIE)
        {
          // produce 8 proteins at a time
          for(i=0;i<8;i++)
{
             NUMBERPROTEINS++;
             add_Protein_agent(NUMBERPROTEINS,1000.0,0,0,1,X,Y);
          }
        }
      }
    }
    return 0;
}

int Source_filewrite()
{
  FILE *file;
  char data[1000];
  char * location="proteindist";
  int protein_dist[100];
  int i;
  int pd=0;

  for(i=0;i<100;i++)
  {
    protein_dist[i]=0;
  }
  //produce a particular text file
  sprintf(data, "%s.txt",location);
  if((file = fopen(data, "a"))==NULL)
  {
    printf("Error: cannot open file '%s' for writing\n", data);
    exit(0);
  }

  proteinnumber_compartment_message=
   get_first_proteinnumber_compartment_message();
  while(proteinnumber_compartment_message)
  {
    pd=proteinnumber_compartment_message->compartment_id-1;
    protein_dist[pd]=proteinnumber_compartment_message->production;

    proteinnumber_compartment_message=
                       get_next_proteinnumber_compartment_message
                                 (proteinnumber_compartment_message);
  }

  sprintf(data, "%d", TIMECOUNTER);
  fputs(data, file);
  for(i=0;i<100;i++)
  {
    fputs(" ", file);
    sprintf(data, "%d", protein_dist[i]);
    fputs(data, file);
  }
```

```
  fputs("\n", file);
  (void)fclose(file);
  return 0;
}
```

7.3.3   Find Optimum Model Settings

To ensure a correct comparison of the two techniques, both models were simulated with identical conditions and data were collected and analyzed. The experience in both simulation techniques was compared across a number of factors like simulation time, the memory size needed and tools used (Table 7.1).

Analyzing the time to actually write the models can be arguable, depending on the experience of the programmers. A programmer with little prior knowledge of agent-based modeling may take more than a month to get accustomed to agents over the platforms. This also includes an installation and learning time for the actual agent-based platforms. For stochastic simulation, MATLAB specializes in mathematical function writing, and would be relatively easier to grasp than a different agent-based modeling framework.

Both models produce results in different ways. Agent-based models produce results as time snapshots for agent conditions at these times. MATLAB can produce concentration gradients, to see the overall system behavior at different times.

The global values can be another deciding factor in how the results look in the end. These need to be tested with multiple runs, to find optimum conditions for the simulation to give results which match closest to real data. Agent-based model results can also use averages over a number of simulations runs, to compensate for the random nature of the agents inherent in the models.

Figure 7.7 shows the intensity plots of the protein distribution across the embryo during the simulation. The figure shows plots (from left to right) deterministic, stochastic and agent-based models. Another representation is shown in Figure 7.8, where peaks of molecule numbers in compartments are compared.

The results show the decay rate being too high in the agent model, where the protein agents die before reaching the last end of the embryo cell (last compartment). Therefore this needs to be reduced in order to match the stochastic behavior. Therefore, using the same initial conditions from the stochastic model, the results are not able to be replicated in the agent-based simulations. Figure 7.9 shows the missing data points in the resulting figures when both models use the same initial setting. The agent-based model is then simulated multiple times with different global conditions to find the best set of values that will produce results closer to the stochastic model (Table 7.2).

TABLE 7.1: Comparing building simulations in MATLAB and FLAME.

Objective	Stochastic	Agents
Total simulation time	200 min for one realization. Total time step is around 3 x 106 (stochastically).	12000 time steps.
Actual simulation time	CPU time: 795.02 seconds.	5 hours.
Memory usage	Approx 420 MB .	Approx 30 GB.
Results format produced	3.1×106 by 100 matrix in MATLAB.	120,000 xml files which are later parsed to produce excel sheets to plot graphs.
Model writing time	1 week. Understanding Gillespie Algorithm and implementation in MATLAB.	1 month, involves understanding the model description and converting to what happens in one iteration.
Global values which can easily be changed	All decay, production and diffusion rates highlighted in starting conditions.	All decay, production and diffusion rates highlighted in starting conditions.
Simulation tool used	MATLAB.	FLAME serial version run on a MAC laptop.
Results measured	The results have measured every minute according to all the compartments as shown in Figure 7.7. In Figure 7.8, the protein distributed at $t = 60$ minutes (9.3×10^5 iteration), $t = 100$ min ($1.6 \times 10^6 iteration$), $t = 144$ min (2.2×10^6 iteration), $t = 180$ min ($2.8 \times 10^6 iteration$), $t = 200$ min (3.1×10^6 iteration).	As number of proteins per time step across compartments, and protein distributions at times 60 min (3600 iteration step (it)), 100 min (6000 it), 144 min (8640 it), 180 min (10800 it), 200 min (12000 it).
Average over runs	One realization was taken. The averaged stochastic model is shown by PDE in [120].	Model was run 20 times and the average was taken.

Due to the large number of cases, a minimum square distance was used to calculate the error rate between the results of each of the cases with the stochastic results, shown in Figure 7.10. The best case which was able to

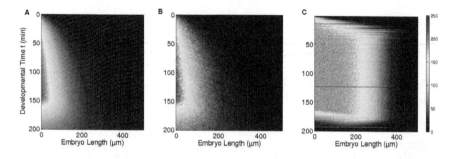

FIGURE 7.7: Bicoid concentration profiles jointly in A-P axis and developmental time, shows a deterministic model output as an average value of stochastic model (A), to one stochastic simulation (B) and the results of one agent-based model run (C).

TABLE 7.2: Different initial value setting for the Bicoid ABM.

Global Value	Range of Values, Interval	Best-Case Found (Case 205)
Source decay rate	[0.01-0.05], 0.1	0.03
Protein decay rate	[0.01-0.05], 0.1	0.01
Probability of protein to move right	[0.1-0.5], 0.1	0.2
Probability of protein to move left	[0.1-0.5], 0.1	0.3
Source production rate	[0.2-1.0], 0.1	0.7

replicate the stochastic conditions was then chosen and shown in Figure 7.11. However, the agent results are still not able to produce a one-to-one mapping of the outputs.

7.4   Output Files for Analysis

To plot the molecule concentration, a particular text file was needed which documented the protein distribution per time. To prevent further code, C functions could be added as agent functions to output the memory variable as a separate text file during the simulation run. This prevents modelers from analyzing the XML result files, and process only one file.

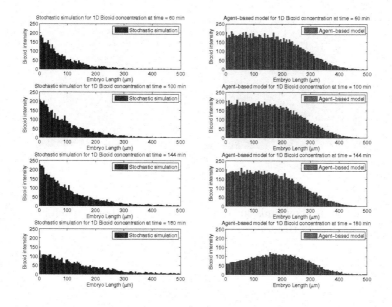

FIGURE 7.8: One realization of stochastic simulation using Gillespie Algorithm at different time points: 60 (A), 100 (B), 144 (C) and 180 (D) min. Blue histograms show number of Bicoid molecules along anterior and posterior axis in embryo. Red lines show average amount of molecules from deterministic reaction diffusion model. Bicoid intensity at 144 min (C) is the peak stage and will degrade after mRNA regulation. Red histograms show number of Bicoid molecules along anterior and posterior axis in embryo resulting from average 20 runs of the agent-based model simulation.

```
//create a file for output
sprintf(data, "%s.txt",location);
if((file = fopen(data, "a"))==NULL)
{
  printf("Error: cannot open file '%s' for writing\n", data);
  exit(0);
}

//start adding data to file from agent memory
sprintf(data, "%d", TIMECOUNTER);
fputs(data, file);
for(i=0;i<100;i++)
{
  fputs(" ", file);
```

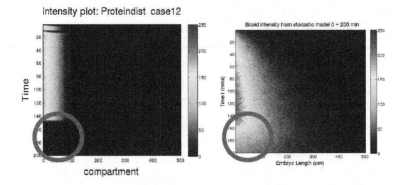

FIGURE 7.9: The agent-based modeling simulation result with stochastic model. The circle shows missing data points in agent-based results using same initial settings in both models.

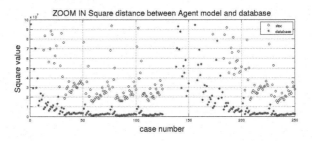

FIGURE 7.10: Zoom in to find shortest possible error between simulated results in agent-based, stochastic and original datasets.

```
  sprintf(data, "%d", protein_dist[i]);
  fputs(data, file);
}
fputs("\n", file);

//close the file
(void)fclose(file);
```

Researchers have compared modeling techniques, such as Norling's technique [143] comparing a systems dynamics and an agent-based model of a food web evolution. In this experiment, a few points can be considered.

- Modelers can discover new details about the model. In equation models, because equations collectively represent agent function as one programming code, modeler is robbed with this opportunity to find new behaviors as a result of this analysis.

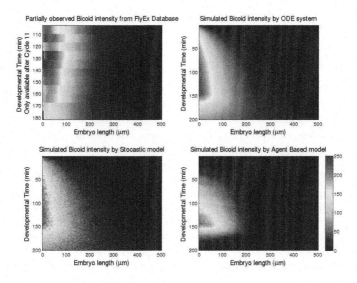

FIGURE 7.11: Using shortest possible error between the simulated results.

- Global values can influence the results produced. Global values can be changed dynamically during the course of the model simulation. This can essentially be done in both kinds of models depending on how the models are written.

- Starting conditions of the model can have an effect on the model results. This is seen in both approaches.

- Introducing dynamic inputs to the model. In an agent-based model, dynamic agents can be introduced which get activated or influence the progression of results. This can easily be programmed by having an agent added which performs certain activities at certain time steps. This would however be tedious to be programmed in an equation model as complicated nested for-loops may need to be added to the model to allow this. This involves very little changes in an agent-based model.

- Increasing complexity. Further complexity can be easily introduced in agent-based models by adding agents and new functions. In equation models, this would require rewriting of the equations and the source code.

- Directed behavior. Agents are autonomous, goal-directed and sociable elements. The decisions they make are based on bounded rationality which means that each agent would have a sphere of influence which allows proximity to be checked before making decisions. In equation models this concept is not present. Here a list is traversed and everything in the list is acted upon in the same manner. In agents, the messages in the sphere of influence may vary allowing agents to display different behaviors depending on where they are located. This is particularly useful when modeling realistic biological models.

- Heterogeneous populations. Different agents who differ in memory can be introduced together in the same simulation. This can produce more interesting results as it brings heterogeneity and how agent internal characteristics can influence results. This cannot be done in equation models as these assume a homogeneous population.

- Bounded rationality. Agents will act depending on their surroundings producing emergent phenomena. This cannot be programmed in an equation model.

- Different scenarios during simulations. Easily different conditions can be introduced to test the model across various conditions. This would not require doing any changes to the agent-based models. Simulations can be stopped halfway, conditions can be changed or new agents can be introduced at adhoc and then simulations can be preceded.

- Large amount of data produced. This is a problematic task to analyze large amounts of data being produced by agent-based models as compared to equation models. Sometimes it is good to find patterns which may not have been thought of previously, but this can be a cumbersome task and may require additional intelligent data mining algorithms at a later stage.

However, it largely depends on the research questions being investigated when a model is being written. In all cases there is a learning curve for biologists and computer scientists to understand which to use and why.

7.5 Modeling Pharaoh's Ants (*Monomorium pharaonis*)

Pharaoh's ant is a 2-mm monomorphic pest ant forming colonies with less than 2500 workers. These ants have poor vision, making them wholly reliant on the pheromones deposited for path directions. Unlike any other ant species, they deposit trail pheromones constitutively when outside the nest, forming branching networks of pheromone trails even before the food sources are discovered [23].

In the model, ant agents are characterized by their identity number, nutritional status, current direction and environmental locations. Each ant agent possesses internal memory which would guide them between resource and nest-site locations.

The ant agents exit a nest site when their hunger level drops below a threshold value and they begin searching for food. Agents deposit pheromones in the environment as they walk and sense the presence of pheromone nearby. If a pheromone trail is present they follow the trail, but if multiple pheromones trails are detected, the ant chooses whcih trail to follow based on concentrations found. Once a food source is located, the ant agent eats 0.02 units of food and stops movement.

If there are no pheromones nearby to follow, then the ant agents move randomly, but include a probabilistic turning matrix known as a 'turning kernel'. This 'turning kernel' is derived from an actual video of Pharaoh's ants, where digital analysis or video tracking was used [23].

The body length of a Pharaoh's ant, 2-mm is represented as 2 units in the model. This is depicted by a single step that an ant agent takes simulating a movement of 2-mm. A realistic foraging space for a typical colony was identified as 25-cm to 150-cm. The models were simulated in an environment of 500-mm by 500-mm.

The environment contains a 'nest' agent placed at the centre of the environment and two food source agents. FLAME also supports dynamic creation of agents allowing pheromone agents to be created when required by the ant agents during the simulation, or when an ant agent takes a step.

Figure 7.12 depict the trials generated by the ant agents during the simulation when two food sources are available. The blue trail depicts the ant paths and the yellow depicts the pheromone deposits along the paths.

```
<xmodel version="2">
<name>Find Nest without Pheromone Smell Model</name>
<version>First version</version>
<description>Probability to leave active trail added 0.001</description>
<author>Mesude Bicak</author>
<date>270110</date>
<environment>
  <functionFiles>
    <file>noPheromoneSmell.c</file>
  </functionFiles>
</environment>

<agents>
<xagent>
  <name>Ant</name>
  <memory>
  <variable><type>int</type><name>antID</name></variable>
  <variable><type>double</type><name>antX</name></variable>
  <variable><type>double</type><name>antY</name></variable>
  <variable><type>double</type><name>foodLevel</name></variable>
  <variable><type>int</type><name>isFed</name></variable>
  <variable><type>int</type><name>isInNest</name></variable>
  <variable><type>int</type><name>antDirection</name></variable>
  <variable><type>int</type><name>state</name></variable>
  <variable><type>int</type><name>pheroFound</name></variable>
```

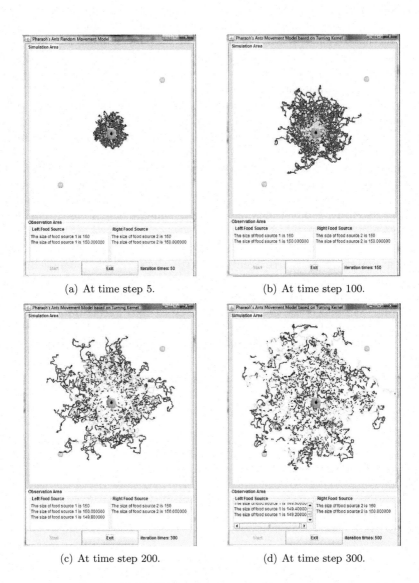

(a) At time step 5.   (b) At time step 100.

(c) At time step 200.   (d) At time step 300.

FIGURE 7.12: Ant simulation.

```
</memory>

<functions>
<function>
<name>updateState</name><description>random coordinates</description>
<currentState>00</currentState>
<nextState>01</nextState>
</function>

<function>
<name>Ant_idle</name><description>idle</description>
<currentState>01</currentState>
<nextState>01a</nextState>
<condition>
<lhs><value>a.state</value></lhs><op>EQ</op><rhs><value>0</value></rhs>
</condition>
</function>

<function>
<name>Ant_idle</name><description>random coordinates for ant</description>
<currentState>01</currentState>
<nextState>01b</nextState>
<condition>
  <lhs><value>a.state</value></lhs><op>NEQ</op><rhs><value>0</value></rhs>
</condition>
</function>

<function>
<name>stayInNest</name><description>random coordinates for ant</description>
<currentState>01a</currentState>
<nextState>04</nextState>
</function>

<function>
<name>walk</name><description>random coordinates for ant</description>
<currentState>01b</currentState>
<nextState>02</nextState>
<condition>
  <lhs><value>a.state</value></lhs><op>NEQ</op><rhs><value>2</value></rhs>
</condition>
<inputs>
  <input><messageName>pheromoneInformation</messageName></input>
  <input><messageName>foodInformation</messageName></input>
</inputs>
</function>

<function>
<name>reinforce</name><description>deposit a pheromone</description>
<currentState>02</currentState>
<nextState>03a</nextState>
<inputs>
  <input><messageName>pheromoneInformation</messageName></input>
</inputs>
<outputs>
  <output><messageName>pheromoneIncreased</messageName></output>
</outputs>
</function>
```

```
<function>
<name>depositPheromone</name>
<currentState>03a</currentState>
<nextState>03</nextState>
<condition>
<lhs><value>a.pheroFound</value></lhs><op>NEQ</op><rhs><value>1</value></rhs>
</condition>
<outputs>
  <output><messageName>newPheromoneInput</messageName></output>
</outputs>
</function>

<function>
<name>idle</name>
<currentState>03a</currentState>
<nextState>03</nextState>
<condition>
  <lhs><value>a.pheroFound</value></lhs><op>EQ</op><rhs><value>1</value></rhs>
</condition>
</function>

<function>
<name>forage</name><description>find a food source</description>
<currentState>03</currentState>
<nextState>04</nextState>
<condition>
  <lhs><value>a.state</value></lhs><op>EQ</op><rhs><value>1</value></rhs>
</condition>
<inputs>
  <input><messageName>foodInformation</messageName></input>
</inputs>
<outputs>
  <output><messageName>foodEaten</messageName></output>
</outputs>
</function>

<function>
<name>forageIdle</name><description>do nothing</description>
<currentState>03</currentState>
<nextState>04</nextState>
<condition>
  <lhs><value>a.state</value></lhs><op>NEQ</op><rhs><value>1</value></rhs>
</condition>
</function>

<function>
<name>decreaseFoodLevel</name><description>increase hunger</description>
<currentState>04</currentState>
<nextState>05</nextState>
</function>

<function>
<name>findNest</name><description>track nest</description>
<currentState>01b</currentState>
<nextState>02</nextState>
<condition>
```

```
      <lhs><value>a.state</value></lhs><op>EQ</op><rhs><value>2</value></rhs>
      </condition>
      <inputs>
        <input><messageName>nestInformation</messageName></input>
        <input><messageName>pheromoneInformation</messageName></input>
      </inputs>
      </function>
      </functions>
  </xagent>

  <xagent>
    <name>Pheromone</name>
    <memory>
      <variable><type>int</type><name>pheromoneID</name></variable>
      <variable><type>double</type><name>life</name></variable>
      <variable><type>double</type><name>pheromoneX</name></variable>
      <variable><type>double</type><name>pheromoneY</name></variable>
    </memory>

    <functions>
    <function>
    <name>pheromoneInformation</name>
    <currentState>00</currentState>
    <nextState>01</nextState>
    <outputs>
      <output><messageName>pheromoneInformation</messageName></output>
    </outputs>
    </function>

    <function>
    <name>decreasePheromoneLife</name>
    <currentState>01</currentState>
    <nextState>02</nextState>
    </function>

    <function>
    <name>increasePheromoneLife</name><description>reinforce pheromone</description>
    <currentState>02</currentState>
    <nextState>03</nextState>
    <inputs><input><messageName>pheromoneIncreased</messageName></input></inputs>
    </function>
    </functions>
  </xagent>

  <xagent>
    <name>Generator</name>
    <memory>
      <variable><type>int</type><name>memoryID</name></variable>
    </memory>

    <functions>
    <function>
    <name>produce</name><description></description>
    <currentState>00</currentState>
    <nextState>01</nextState>
    <inputs>
      <input><messageName>newPheromoneInput</messageName></input>
```

```
    </inputs>
    </function>
    </functions>
</xagent>

<xagent>
  <name>FoodGenerator</name>
  <memory>
    <variable><type>int</type><name>memoryFoodID</name></variable>
  </memory>

  <functions>
  <function>
  <name>produceFood</name><description></description>
  <currentState>00</currentState>
  <nextState>01</nextState>
  <inputs>
    <input><messageName>newFood</messageName></input>
  </inputs>
  </function>
  </functions>
</xagent>

<xagent>
  <name>Nest</name>
  <memory>
    <variable><type>double</type><name>nestX</name></variable>
    <variable><type>double</type><name>nestY</name></variable>
<variable><type>double</type><name>nestRadius</name></variable>
  </memory>

  <functions>
  <function>
  <name>nestInformation</name><description>coordinates of the nest</description>
  <currentState>00</currentState>
  <nextState>01</nextState>
  <outputs>
    <output><messageName>nestInformation</messageName></output>
  </outputs>
  </function>
  </functions>
</xagent>

<xagent>
  <name>Food</name>
  <memory>
    <variable><type>int</type><name>foodID</name></variable>
<variable><type>double</type><name>size</name></variable>
<variable><type>double</type><name>foodX</name></variable>
<variable><type>double</type><name>foodY</name></variable>
<variable><type>double</type><name>radius</name></variable>
  </memory>

  <functions>
  <function>
  <name>foodInformation</name><description>coordinates of the food</description>
  <currentState>00</currentState>
```

```
  <nextState>01</nextState>
  <outputs>
    <output><messageName>foodInformation</messageName></output>
  </outputs>
  </function>

  <function>
  <name>updateFood</name><description>coordinates of the food</description>
  <currentState>01</currentState>
  <nextState>02</nextState>
  <inputs>
    <input><messageName>foodEaten</messageName></input>
  </inputs>
  <outputs>
    <output><messageName>newFood</messageName></output>
  </outputs>
  </function>
  </functions>
</xagent>
</agents>

<messages>
  <message>
    <name>pheromoneInformation</name>
    <description>pheromone deposition</description>
<variables>
  <variable><type>int</type><name>pheromoneID</name></variable>
      <variable><type>double</type><name>pheromoneX</name></variable>
      <variable><type>double</type><name>pheromoneY</name></variable>
      <variable><type>double</type><name>life</name></variable>
      </variables>
  </message>

  <message>
<name>newPheromoneInput</name>
<description>pheromone deposition</description>
<variables>
  <variable><type>double</type><name>pheromoneX</name></variable>
  <variable><type>double</type><name>pheromoneY</name></variable>
</variables>
  </message>

  <message>
<name>foodInformation</name>
<description>food coordinates</description>
<variables>
  <variable><type>int</type><name>foodID</name></variable>
  <variable><type>double</type><name>foodX</name></variable>
  <variable><type>double</type><name>foodY</name></variable>
  <variable><type>double</type><name>size</name></variable>
  <variable><type>double</type><name>radius</name></variable>
</variables>
  </message>

  <message>
    <name>foodEaten</name>
<description>food eaten</description>
```

```
<variables>
  <variable><type>int</type><name>id</name></variable>
  <variable><type>double</type><name>size</name></variable>
</variables>
  </message>

  <message>
    <name>newFood</name>
<description>new food agent</description>
<variables>
  <variable><type>double</type><name>size</name></variable>
  <variable><type>double</type><name>foodX</name></variable>
  <variable><type>double</type><name>foodY</name></variable>
  <variable><type>double</type><name>radius</name></variable>
</variables>
  </message>

  <message>
    <name>pheromoneIncreased</name>
<description>new food agent</description>
<variables>
  <variable><type>int</type><name>pheromoneID</name></variable>
  <variable><type>double</type><name>increase</name></variable>
</variables>
  </message>

  <message>
<name>nestInformation</name>
<description>nest coordinates</description>
<variables>
  <variable><type>double</type><name>nestX</name></variable>
  <variable><type>double</type><name>nestY</name></variable>
  <variable><type>double</type><name>nestRadius</name></variable>
</variables>
  </message>
</messages>
</xmodel>

#include "header.h"
#include "Ant_agent_header.h"
#include "Pheromone_agent_header.h"
#include "Generator_agent_header.h"
#include "Food_agent_header.h"
#include "FoodGenerator_agent_header.h"
#include "Nest_agent_header.h"

//ant size = 2 mm = 6 pixels => 1 mm = 3 pixels
//ant speed = 12.99 pixels/second = 4.33 mm/s
//ant step size = 2 mm = 6 pixels

#define xLeftBorder 20.0
#define xRightBorder 500.0
#define yUpperBorder 20.0
#define yLowerBorder 500.0

#define nestXRightBorder 260.0
#define nestXLeftBorder 240.0
```

```
#define nestYUpperBorder 240.0
#define nestYLowerBorder 260.0

#define antStepSize 2.0
#define antMinPheromoneDetectionUnit 1.0
#define antFoodLevelDecay 0.02
#define antFoodFound 100
#define antPheromoneDepositionUnit 2.0

#define minPheromoneDistance 1.0
#define pheromoneDecay 0.0248

#define stateWalk 1
#define stateFindNest 2
#define stateStayInNest 0

struct Data {
  int maxIndex;
  double information1[1000];
  double information2[1000];
};

struct PheromoneData {
  double pheromoneX;
  double pheromoneY;
  double pheromoneLife;
  int direction;
};

int getDirection(double, double, double, double);

double getDistance(double x, double y, double tx, double ty) {
  return sqrt(pow(tx - x, 2) + pow(ty - y, 2));
}

//Environment: 500, 500
double checkAntPositionY(double antPosition) {
  double newPosition = antPosition;

  if (antPosition < yUpperBorder) {
    newPosition = yUpperBorder;
  }

  if (antPosition > yLowerBorder) {
    newPosition = yLowerBorder;
  }
  return newPosition;
}

double checkAntPositionX(double antPosition) {
  double newPosition = antPosition;

  if (antPosition < xLeftBorder) {
    newPosition = xLeftBorder;
  }

  if (antPosition > xRightBorder) {
```

```
    newPosition = xRightBorder;
  }
  return newPosition;
}

int getDirection(double aX, double aY, double tX, double tY) {
  double tmp = aY;
  aY = tY;
  tY = tmp;

  double vx = tX - aX;
  double vy = tY - aY;

  double length = sqrt((vx * vx) + (vy * vy));
  double alpha = acos(vy / length) * (180.0 / 3.142);

  if (vx < 0) {
    alpha = 360 - alpha;
  }

  int result = (int) round(alpha / 360 * 8);
  if (result == 0) {
    return 8;
  }
  return result;
}

void updatePosition(double* aX, double* aY, int direction) {
  double b = sqrt(antStepSize * antStepSize / 2.0);
  switch (direction) {
  case 8:
    *aY -= antStepSize;
    break;
  case 4:
    *aY += antStepSize;
    break;
  case 1:
    *aX += b;
    *aY -= b;
    break;
  case 2:
    *aX += antStepSize;
    break;
  case 3:
    *aX += b;
    *aY += b;
    break;
  case 5:
    *aX -= b;
    *aY += b;
    break;
  case 6:
    *aX -= antStepSize;
    break;
  case 7:
    *aX -= b;
    *aY -= b;
```

```
      break;
    default:
      break;
    }
}

int getNewDirection(int currentDir, double angle) {
    double vx = 0;
    double vy = 0;

    switch (currentDir) {
    case 8:
      vy = 1;
      break;
    case 4:
      vy = -1;
      break;
    case 1:
      vx = 1;
      vy = 1;
      break;
    case 2:
      vx = 1;
      break;
    case 3:
      vx = 1;
      vy = -1;
      break;
    case 5:
      vx = -1;
      vy = -1;
      break;
    case 6:
      vx = -1;
      break;
    case 7:
      vx = -1;
      vy = 1;
      break;
    default:
      break;
    }

    double length = sqrt((vx * vx) + (vy * vy));
    double alpha = acos(vy / length) * (180.0 / 3.142);

    if (vx < 0) {
      alpha = 360 - alpha;
    }

    double totalAngle = alpha + angle;
    if (totalAngle > 360) {
      totalAngle -= 360;
    }
    if (totalAngle < 0) {
      totalAngle += 360;
    }
```

```
  int result = (int) round(totalAngle / 360 * 8);
  if (result == 0) {
    return 8;
  }
  return result;
}

void detectPheromone(double* pheromoneX, double* pheromoneY) {
  double pheromoneLife = -1;
  double currentX = ANTX;
  double currentY = ANTY;

  int direction = ANTDIRECTION;
  int nextdir = (direction + 1) > 8 ? 1 : direction + 1;
  int prevdir = (direction - 1) < 1 ? 8 : direction - 1;

  pheromoneInformation_message = get_first_pheromoneInformation_message();
  double distanceToPheromone = 0;

  double randomNumber1 = ((double) rand() / ((double) RAND_MAX));

  //leaving trail probability
  if (randomNumber1 <= 0.001){
    pheromoneLife = -1;
    *pheromoneX = -1;
    *pheromoneY = -1;
  }
  else {
    while (pheromoneInformation_message) {
      double pheromoneLocalX = pheromoneInformation_message->pheromoneX;
      double pheromoneLocalY = pheromoneInformation_message->pheromoneY;
      double pheromoneLocalLife = pheromoneInformation_message->life;
      distanceToPheromone = getDistance(pheromoneLocalX,
                        pheromoneLocalY, currentX, currentY);

      if (distanceToPheromone <= antStepSize && pheromoneLocalLife >= 0.2
                    && distanceToPheromone > minPheromoneDistance)
      {
        int newDirection = getDirection(ANTX, ANTY,
                                pheromoneLocalX, pheromoneLocalY);
        if (newDirection == direction || newDirection == nextdir ||
                                newDirection == prevdir) {
        if (pheromoneLife <= pheromoneLocalLife) {
            pheromoneLife = pheromoneLocalLife;
            *pheromoneX = pheromoneLocalX;
            *pheromoneY = pheromoneLocalY;
        }
  }
}
    }
  pheromoneInformation_message = get_next_pheromoneInformation_message
                                        (pheromoneInformation_message);
  }
  }
}

int turningKernel() {
```

```
  //Check if moving or stationary
  double randomNumber1 = ((double) rand() / ((double) RAND_MAX));
  double currentX = ANTX;
  double currentY = ANTY;
  double newX = currentX;
  double newY = currentY;
  int direction = ANTDIRECTION;
  int newDirection = ANTDIRECTION;

  //Moving: 66%
  if (randomNumber1 <= 0.66) {
    double randomNumber2 = ((double) rand() / (double) RAND_MAX);
    //DIR 8 = 0 degrees ahead = 41%
    if (randomNumber2 <= 0.41) {
newDirection = getNewDirection(direction, 0);
//DIR 1 = 45 degrees clockwise = 27%
    }
    else if (randomNumber2 <= 0.68) {
newDirection = getNewDirection(direction, 45);
//DIR 7 = 45 degrees anti-clockwise = 19.0%
    }
    else if (randomNumber2 <= 0.87) {
      newDirection = getNewDirection(direction, -45);
//DIR 2 = 90 degrees clockwise = 5.4%
    }
    else if (randomNumber2 <= 0.924) {
      newDirection = getNewDirection(direction, 90);
      //DIR 6 = 90 degrees anti-clockwise = 5.4%
    } else if (randomNumber2 <= 0.978) {
      newDirection = getNewDirection(direction, -90);
      //DIR 3 = 135 degrees clockwise = 2.7%
    }
    else {
      newDirection = getNewDirection(direction, 135);
    }
    updatePosition(&newX, &newY, newDirection);
    ANTY = checkAntPositionY(newY);
    ANTX = checkAntPositionX(newX);
    ANTDIRECTION = newDirection;
    //checkAntWalkThroughNest(ANTX, ANTY);
  }
  else {
    //Stationary = 33%
  }
  return 0;
}

int updateState(){
  if ((ISINNEST == 1 && FOODLEVEL < 1.0) || (ISFED == 0
                                      && FOODLEVEL < 1.0)) {
    STATE = stateWalk;
    //printf("walk & forage!\n");
  }
  else if (ISFED == 1 && ISINNEST != 1) {
    STATE = stateFindNest;
    //printf("find nest!\n");
  }
```

```
  else if (ISINNEST == 1 && FOODLEVEL >= 1.0) {
    STATE = stateStayInNest;
//printf("stay in nest!\n");
  }
  else {
    //printf("this should never happen!\n");
  }
  return 0;
}

int stayInNest(){
  //printf("stay in nest called: ant %d\n", ANTID);
  return 0;
}

int forageIdle(){
  //printf("forage idle called: ant %d\n", ANTID);
  return 0;
}

int depositIdle(){
  //printf("deposit idle called: ant %d\n", ANTID);
  return 0;
}

//Ant agent
//walk based on turning kernel
int walk() {
  double pheromoneX = -1;
  double pheromoneY = -1;
  ISINNEST = 0;
  detectPheromone(&pheromoneX, &pheromoneY);

  //no pheromones nearby
  if (pheromoneX == -1 && pheromoneY == -1)
  {
    turningKernel();
  } else {
    //pheromone found
    double currentX = ANTX;
    double currentY = ANTY;
    int direction = getDirection(ANTX, ANTY, pheromoneX, pheromoneY);
    updatePosition(&currentX, &currentY, direction);
    ANTX = checkAntPositionX(currentX);
    ANTY = checkAntPositionY(currentY);
    ANTDIRECTION = direction;
  }
  return 0;
}

//Ant agent
//decrease foodLevel at each iteration
int decreaseFoodLevel() {
  //printf("in decrease food level\n");
  FOODLEVEL = FOODLEVEL - 0.02;

  if (FOODLEVEL <= 0) {
```

```
    FOODLEVEL = 0;
  }

  if (FOODLEVEL <1.0 && ISFED!=0){
    ISFED = 0;
  }
  PHEROFOUND = 0;
  return 0;
}


//Ant agent
//find a food source and eat 0.2 units
int forage() {
  int localId;
  foodInformation_message = get_first_foodInformation_message();
  while (foodInformation_message) {
    double xFood = foodInformation_message->foodX;
    double yFood = foodInformation_message->foodY;
    double foodRadius = foodInformation_message->radius;
    double distanceToFood = getDistance(xFood, yFood, ANTX, ANTY);

    if (distanceToFood <= foodRadius + 2) {
      double foodSize = foodInformation_message->size;
      localId = foodInformation_message->foodID;
      FOODLEVEL = FOODLEVEL + 100;
      ISFED = 1;
      add_foodEaten_message(localId, (foodSize - 0.2));
      int direction = ANTDIRECTION;
      ANTDIRECTION = getNewDirection(direction, 180);
      return 0;
    }
    foodInformation_message = get_next_foodInformation_message
                              (foodInformation_message);
  }
  return 0;
}


//FoodGenerator agent
//creates a new food agent dynamically by checking newFood_message
int produceFood() {
  newFood_message = get_first_newFood_message();
  while (newFood_message) {
    double x = newFood_message->foodX;
    double y = newFood_message->foodY;
    double foodSize = newFood_message->size;
    double foodRadius = newFood_message->radius;
    MEMORYFOODID++;
    add_Food_agent(MEMORYFOODID, foodSize, x, y, foodRadius);
    newFood_message = get_next_newFood_message(newFood_message);
  }
  return 0;
}


//Food agent
//if one of the food sources is depleted, create newFood_message
//which is passed to produceFood() function for FoodGenerator agent
int updateFood() {
```

```
    foodEaten_message = get_first_foodEaten_message();
    while (foodEaten_message) {
      int localID = foodEaten_message->id;
      if (localID == FOODID) {
        SIZE = foodEaten_message->size;
        if (SIZE <= 0.2) {
          //printf("creating new food\n");
          double xPosition = rand() / (double) (RAND_MAX) * 480;
          double yPosition = rand() / (double) (RAND_MAX) * 480;
          SIZE = (rand() / (double) (RAND_MAX) * 100) + 1;

          if (SIZE > 0 && SIZE < 5) {
            RADIUS = 5.0;
          }
          else if (SIZE >= 5 && SIZE < 10) {
            RADIUS = 6.0;
          }
          else if (SIZE >= 10 && SIZE < 20) {
            RADIUS = 7.0;
          }
          else if (SIZE >= 20 && SIZE < 30) {
            RADIUS = 8.0;
          }
          else if (SIZE >= 30 && SIZE < 40) {
            RADIUS = 9.0;
          } else if (SIZE >= 40 && SIZE < 50) {
            RADIUS = 10.0;
          } else if (SIZE >= 50 && SIZE < 60) {
            RADIUS = 11.0;
          } else if (SIZE >= 60 && SIZE < 70) {
            RADIUS = 12.0;
          } else if (SIZE >= 70 && SIZE < 80) {
            RADIUS = 13.0;
          } else if (SIZE >= 80 && SIZE < 90) {
            RADIUS = 14.0;
          } else if (SIZE >= 90) {
            RADIUS = 15.0;
          }
          add_newFood_message(SIZE, xPosition, yPosition, RADIUS);
          return 1;
        }
      }
      foodEaten_message = get_next_foodEaten_message(foodEaten_message);
    }
    return 0;
}

//Food agent
//add foodInfo
int foodInformation() {
  if (SIZE > 0 && SIZE < 5) {
    RADIUS = 5.0;
  } else if (SIZE >= 5 && SIZE < 10) {
    RADIUS = 6.0;
  } else if (SIZE >= 10 && SIZE < 20) {
    RADIUS = 7.0;
  } else if (SIZE >= 20 && SIZE < 30) {
```

```
      RADIUS = 8.0;
    } else if (SIZE >= 30 && SIZE < 40) {
      RADIUS = 9.0;
    } else if (SIZE >= 40 && SIZE < 50) {
      RADIUS = 10.0;
    } else if (SIZE >= 50 && SIZE < 60) {
      RADIUS = 11.0;
    } else if (SIZE >= 60 && SIZE < 70) {
      RADIUS = 12.0;
    } else if (SIZE >= 70 && SIZE < 80) {
      RADIUS = 13.0;
    } else if (SIZE >= 80 && SIZE < 90) {
      RADIUS = 14.0;
    } else if (SIZE >= 90) {
      RADIUS = 15.0;
    }
    add_foodInformation_message(FOODID, FOODX, FOODY, SIZE, RADIUS);
    return 0;
}


//Nest agent
//add nestInfo
int nestInformation() {
    add_nestInformation_message(NESTX, NESTY, NESTRADIUS);
    return 0;
}


//Pheromone agent
//add pheromoneInfo
int pheromoneInformation() {
    add_pheromoneInformation_message(PHEROMONEID, PHEROMONEX, PHEROMONEY, LIFE);
    return 0;
}


int Ant_idle()
{
    return 0;
}


//Ant agent
//deposit a pheromone at each step
//if a pheromone was previously deposited at a particular coordinate,
  increase pheromone level
//if not create a new one
int reinforce()
{
    double distance=0.0;
    pheromoneInformation_message = get_first_pheromoneInformation_message();
    while (pheromoneInformation_message)
    {
      if(PHEROFOUND == 0)
      {
        distance = getDistance(ANTX, ANTY, pheromoneInformation_message->
                   pheromoneX, pheromoneInformation_message->pheromoneY);
        if (distance > 0.3 && distance <= minPheromoneDistance)
        {
          PHEROFOUND = 1;
```

```
      add_pheromoneIncreased_message(pheromoneInformation_message->
                      pheromoneID, antPheromoneDepositionUnit); //0.0496
      }
    }
    pheromoneInformation_message = get_next_pheromoneInformation_message
                                  (pheromoneInformation_message);
  }
  return 0;
}


int depositPheromone()
{
  //printf("phero not found, create one at ant coordinates %f %f\n" ,ANTX,ANTY);
  add_newPheromoneInput_message(ANTX, ANTY);
  return 0;
}


//Pheromone agent
//decrease pheromoneLife at each iteration
int decreasePheromoneLife() {
  if (LIFE >= 0.02) {
    LIFE = LIFE - (pheromoneDecay * LIFE);
  }
  if (LIFE < 0.02) { //pheromone evaporated
    return 1;
  }
  return 0;
}


//Pheromone agent
//increases the life of a previously deposited pheromone at a specific location
int increasePheromoneLife() {
  pheromoneIncreased_message = get_first_pheromoneIncreased_message();
  while (pheromoneIncreased_message) {
    int localPheromoneID = pheromoneIncreased_message->pheromoneID;
    if (localPheromoneID == PHEROMONEID) {
      LIFE = LIFE + pheromoneIncreased_message->increase;
    }
    pheromoneIncreased_message = get_next_pheromoneIncreased_message
                                  (pheromoneIncreased_message);
  }
  return 0;
}


//Generator agent
//produces a pheromone agent dynamically
//prevents duplicate deposition
int produce() {
  newPheromoneInput_message = get_first_newPheromoneInput_message();
  struct Data data;
  data.maxIndex = 0;
  while (newPheromoneInput_message)
  {
    double x = newPheromoneInput_message->pheromoneX;
    double y = newPheromoneInput_message->pheromoneY;
    int found = 0;
    for (int i = 0; i < data.maxIndex; i++)
```

```
  {
    if((data.information1[i] == x) && (data.information2[i] == y))
    {
      found = 1;
    }
  }
  if (found == 0)
  {
    data.information1[data.maxIndex] = x;
    data.information2[data.maxIndex] = y;
    data.maxIndex = data.maxIndex + 1;
  }
  newPheromoneInput_message = get_next_newPheromoneInput_message
                                    (newPheromoneInput_message);
}
for (int i = 0; i < data.maxIndex; i++)
{
  MEMORYID++;
  // printf("New agent");
  add_Pheromone_agent(MEMORYID, antPheromoneDepositionUnit,
                    data.information1[i], data.information2[i]);
}
return 0;
}


//Ant agent
//after eating food, ant agent goes back to nest
int findNest() {
  int pheromoneFound = 0;
  int epFound = 0;
  double currentX = ANTX;
  double currentY = ANTY;
  struct PheromoneData p;
  p.pheromoneX = -1;
  p.pheromoneY = -1;
  p.pheromoneLife = -1;
  p.direction = ANTDIRECTION;
  struct PheromoneData ep;
  ep.pheromoneX = -1;
  ep.pheromoneY = -1;
  ep.pheromoneLife = -1;
  ep.direction = ANTDIRECTION;
  int direction = ANTDIRECTION;
  int nextdir = (direction + 1) > 8 ? 1 : direction + 1;
  int prevdir = (direction - 1) < 1 ? 8 : direction - 1;
  int next2dir = (direction + 2) > 8 ? 1 : direction + 2;
  int prev2dir = (direction - 2) < 1 ? 8 : direction - 2;
  double distanceToPheromone = 0;

  nestInformation_message = get_first_nestInformation_message();
    while (nestInformation_message) {
      double xNest = nestInformation_message->nestX;
      double yNest = nestInformation_message->nestY;
      double nestRadius = nestInformation_message->nestRadius;
      double distanceToNest = getDistance(xNest, yNest, currentX, currentY);
      if (distanceToNest <= nestRadius + 2) {
        ISINNEST = 1;
```

```
        return 0;
  }
  nestInformation_message = get_next_nestInformation_message
                                    (nestInformation_message);
}
  pheromoneInformation_message = get_first_pheromoneInformation_message();
  while (pheromoneInformation_message) {
    double pheromoneLocalX = pheromoneInformation_message->pheromoneX;
    double pheromoneLocalY = pheromoneInformation_message->pheromoneY;
    double pheromoneLocalLife = pheromoneInformation_message->life;
    distanceToPheromone = getDistance(pheromoneLocalX, pheromoneLocalY,
                                      currentX, currentY);

if (distanceToPheromone <= antStepSize && pheromoneLocalLife > 0.2
                  && distanceToPheromone > minPheromoneDistance) {
      int newDirection = getDirection(ANTX, ANTY, pheromoneLocalX, pheromoneLocalY);
        if (newDirection == direction || newDirection == nextdir
                                        || newDirection == prevdir) {
    if (p.pheromoneLife <= pheromoneLocalLife) {
    p.pheromoneLife = pheromoneLocalLife;
            p.pheromoneX = pheromoneLocalX;
            p.pheromoneY = pheromoneLocalY;
            p.direction = getDirection(ANTX, ANTY, p.pheromoneX, p.pheromoneY);
            pheromoneFound = 1;
          }
        }
        else if (newDirection == next2dir || newDirection == prev2dir) {
          if (ep.pheromoneLife <= pheromoneLocalLife) {
            ep.pheromoneLife = pheromoneLocalLife;
            ep.pheromoneX = pheromoneLocalX;
            ep.pheromoneY = pheromoneLocalY;
            ep.direction = getDirection(ANTX, ANTY, ep.pheromoneX, ep.pheromoneY);
            epFound = 1;
          }
        }
      }
      pheromoneInformation_message = get_next_pheromoneInformation_message
                                    (pheromoneInformation_message);
    }
    if (pheromoneFound == 1) {
      direction = p.direction;
      updatePosition(&currentX, &currentY, direction);
      ANTX = checkAntPositionX(currentX);
      ANTY = checkAntPositionY(currentY);
      ANTDIRECTION = direction;
      //checkAntWalkThroughNest(ANTX, ANTY);
    } else if (epFound == 1) {
      direction = ep.direction;
      //printf("after pheromone: direction is %d\n", direction);
      updatePosition(&currentX, &currentY, direction);
      ANTX = checkAntPositionX(currentX);
      ANTY = checkAntPositionY(currentY);
      ANTDIRECTION = direction;
      //checkAntWalkThroughNest(ANTX, ANTY);
    } else {
      turningKernel();
    }
  }
```

```
  return 0;
}
```

<!-- starting file 0.xml-->
<states>
<itno>0</itno>
<xagent>
  <name>Ant</name>
  <antID>1</antID>
  <antX>247.92854316936237</antX>
  <antY>254.99745892924577</antY>
  <foodLevel>0</foodLevel>
  <isFed>0</isFed>
  <isInNest>1</isInNest>
  <antDirection>4</antDirection>
  <state>0</state>
  <pheroFound>0</pheroFound>
</xagent>
<xagent>
  <name>Ant</name>
  <antID>2</antID>
  <antX>247.71740888841424</antX>
  <antY>253.6156297463214</antY>
  <foodLevel>0</foodLevel>
  <isFed>0</isFed>
  <isInNest>1</isInNest>
  <antDirection>2</antDirection>
  <state>0</state>
  <pheroFound>0</pheroFound>
</xagent>
<xagent>
  <name>Pheromone</name>
  <pheromoneID>1</pheromoneID>
  <life>0</life>
  <pheromoneX>0</pheromoneX>
  <pheromoneY>0</pheromoneY>
</xagent>
<xagent>
  <name>Generator</name>
  <memoryID>1</memoryID>
</xagent>
<xagent>
  <name>FoodGenerator</name>
  <memoryFoodID>2</memoryFoodID>
</xagent>
<xagent>

```
  <name>Nest</name>
  <nestX>250</nestX>
  <nestY>250</nestY>
  <nestRadius>10</nestRadius>
</xagent>
<xagent>
  <name>Food</name>
  <foodID>1</foodID>
  <size>100</size>
  <foodX>100</foodX>
  <foodY>400</foodY>
  <radius>15</radius>
</xagent>
<xagent>
  <name>Food</name>
  <foodID>2</foodID>
  <size>5</size>
  <foodX>300</foodX>
  <foodY>200</foodY>
  <radius>5</radius>
</xagent>
</states>
```

7.6   Model Drug Delivery for Cancer Treatment

Curing cancer is a game of time and drugs where drugs are introduced at specific times to help kill cancer cells before they mutate. Cancer cells eventually develop resistance to the administered drugs during the treatments. This is a concrete factor in limiting treatment, causing the patient to move to a point of no return when no more chemotherapy is effective. Individual cells develop changes in their DNA (known as 'mutations') that change them such that their growth is left unaffected by specific drugs. Through a computer model that allows introducing alternative drugs, in a virtual environments on computer-generated cancer cells, clinicians can find best drug combinations to reduce the development of drug resistance, increasing the patient chances to survive. This project involves working with breast cancer patients but can be extended to other types of cancers in the future.

Drug resistance is the main reason for current failures in cancer treatments when no more chemotherapy is helpful for the patient. Clinicians use a combination of drugs (drug A and drug B) introducing them at different times of the treatment in order to help kill all cancer cells in the affected tissues

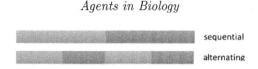

FIGURE 7.13: Sequential trails for drug therapy.

such as shown in Figure 7.13. There is a need to understand what is the best cost-affective pattern which can help kill off all cancer cells before they all mutate to develop resistances to all kinds of drugs. Using agent-based modeling and working with clinicians, a simulation was developed to build tools to allow clinicians to test their theories in controlled virtual environments. These tools can help save on costs of the drugs and materials and also prevent the delayed wait times for when the experiments are conducted in petri dishes on real cancer-affected tissues. By simulating the cell behavior, we can quickly find approximate best combinations of drugs allowing the clinicians to zero in on the combinations they can test out in laboratories, testing their simulated hypothesis saving on time and experiments. Below is the overall simulation behavior:

- At start of iteration: Generate cells on random with 2 state mutation categories (0 - neutral, 1 relieves pressure of drug A, 2 relieves pressure of drug B).

- During simulation:

    - Cells continue to divide based on growth rate/division rate.

    - Introduce drug A cells into simulation at specified intervals.

    - If drug A close by, cells with mutation state 1 will fight and die, or if neutral: reduce growth rate of cells, or if state 1: kill cell, apply decaying function for cell to die.

    - If drug A kills certain cells close by, remove that part of drug A from scenario.

    - Introduce drug B in scenario. Repeat process with drug B.

    - Save data at each time step.

The effect of alternating drugs is simple, but judging from the initial models it can prove highly successful, because the subset of cells that developed drug resistance to the one drug are destroyed when drugs are alternated, and vice versa. The alternating drug strategy therefore reduces the risk of the cancer developing dual resistance to both drugs, because the effective population size for this mutation to develop in is smaller. It is this dual (or multiple resistance to > 2 drugs) resistance that will ultimately render the cancer untreatable. The effective population size of cells that have resistance to one of the drugs is crucial because this determines the chance that dual resistance can evolve. With this project, the authors developed the agent-based model to include competition between cancer cells for resources which will add increased complexity, using population level modeling. In this model, they showed that a sequential way of treating cancer is more prone to encourage the development of drug resistance. The current clinical strategy is mainly sequential, first treating with one drug and using sequential drugs to battle any recurrence or when treatment on earlier drugs was unsuccessful. Many drugs can usually not be provided as a cocktail because this will pose severe side effects. Currently, it is the aim to extend these models to include interactions between cells in terms of position within the cancer and competition for oxygen and energy.

```xml
<?xml version="1.0" encoding="ISO-8859-1"?>
<xmodel version="1">
<name>Cancer Drug Model</name>
<date>150714</date>

<!--********** Environment values and functions ********-->
<environment>
  <functionFiles>
    <file>cell_functions.c</file>
    <file>druga_functions.c</file>
    <file>drugb_functions.c</file>
<file>generatedrug_functions.c</file>
<file>library_functions.c</file>
  </functionFiles>
</environment>

<agents>
  <xagent>
    <name>Cell</name>
    <memory>
      <variable><type>int</type><name>myid</name></variable>
      <variable><type>int</type><name>mcat</name></variable>
      <variable><type>double</type><name>xpos</name></variable>
      <variable><type>double</type><name>ypos</name></variable>
      <variable><type>double</type><name>zpos</name></variable>
  <variable><type>double</type><name>clife</name></variable>
</memory>

    <functions>
    <function>
      <name>cell_here</name>
      <currentState>0a</currentState>
      <nextState>00</nextState>
      <outputs>
        <output><messageName>im_here</messageName></output>
      </outputs>
    </function>

    <function>
      <name>cell_divide</name><description>divide on division rate</description>
      <currentState>00</currentState>
      <nextState>01</nextState>
```

```
    <inputs>
      <input><messageName>cell_tot_msg</messageName></input>
    </inputs>
  </function>

  <function>
    <name>cell_check_drugs</name><description>affected by drug</description>
    <currentState>01</currentState>
    <nextState>02</nextState>
    <inputs>
      <input><messageName>drug_a_information</messageName></input>
      <input><messageName>drug_b_information</messageName></input>
    </inputs>
    <outputs>
      <output><messageName>drug_used</messageName></output>
    </outputs>
  </function>

  <function>
    <name>cell_decay</name>
    <currentState>02</currentState>
    <nextState>03</nextState>
  </function>
  </functions>
</xagent>

<xagent>
  <name>DrugA</name>
  <memory>
    <variable><type>int</type><name>myid</name></variable>
    <variable><type>double</type><name>xpos</name></variable>
    <variable><type>double</type><name>ypos</name></variable>
    <variable><type>double</type><name>zpos</name></variable>
    <variable><type>double</type><name>alife</name></variable>
  </memory>

  <functions>
    <function>
    <name>drug_a_location</name>
    <currentState>00</currentState>
    <nextState>01</nextState>
    <outputs>
      <output><messageName>drug_a_information</messageName></output>
    </outputs>
    </function>

    <function>
    <name>drug_a_used</name>
    <currentState>01</currentState>
    <nextState>02</nextState>
    <inputs><input><messageName>drug_used</messageName></input></inputs>
    </function>

    <function>
    <name>drug_a_decay</name>
    <currentState>02</currentState>
    <nextState>03</nextState>
```

```
      </function>
  </functions>
  </xagent>

<xagent>
  <name>DrugB</name>
  <memory>
    <variable><type>int</type><name>myid</name></variable>
    <variable><type>double</type><name>xpos</name></variable>
    <variable><type>double</type><name>ypos</name></variable>
    <variable><type>double</type><name>zpos</name></variable>
    <variable><type>double</type><name>blife</name></variable>
  </memory>

  <functions>
    <function>
      <name>drug_b_location</name>
      <currentState>00</currentState>
      <nextState>01</nextState>
      <outputs>
        <output><messageName>drug_b_information</messageName></output>
      </outputs>
    </function>

    <function>
      <name>drug_b_used</name>
      <currentState>01</currentState>
      <nextState>02</nextState>
      <inputs>
        <input><messageName>drug_used</messageName></input>
      </inputs>
    </function>

    <function>
      <name>drug_b_decay</name>
      <currentState>02</currentState>
      <nextState>03</nextState>
    </function>
  </functions>
  </xagent>

<xagent>
  <name>DrugGenerator</name>
  <memory>
    <variable><type>int</type><name>myid</name></variable>
    <variable><type>int</type><name>aid</name></variable>
    <variable><type>int</type><name>bid</name></variable>
    <variable><type>int</type><name>time_count</name></variable>
    <variable><type>int</type><name>total_cells</name></variable>
    <variable><type>double</type><name>xpos</name></variable>
    <variable><type>double</type><name>ypos</name></variable>
    <variable><type>double</type><name>zpos</name></variable>
  </memory>

  <functions>
    <function>
      <name>Generate_drug</name>
```

```
            <currentState>00</currentState>
            <nextState>01</nextState>
          </function>

          <function>
            <name>Count_cells</name>
            <currentState>01</currentState>
            <nextState>02</nextState>
            <inputs>
                <input><messageName>im_here</messageName></input>
            </inputs>
            <outputs>
                <output><messageName>cell_tot_msg</messageName></output>
            </outputs>
          </function>
        </functions>
    </xagent>
</agents>

<messages>
  <message>
    <name>drug_a_information</name>
    <variables>
      <variable><type>int</type><name>drugid</name></variable>
      <variable><type>double</type><name>myx</name></variable>
      <variable><type>double</type><name>myy</name></variable>
    </variables>
  </message>

  <message>
    <name>drug_b_information</name>
    <variables>
      <variable><type>int</type><name>drugid</name></variable>
      <variable><type>double</type><name>myx</name></variable>
      <variable><type>double</type><name>myy</name></variable>
    </variables>
  </message>

  <message>
    <name>drug_used</name>
    <variables>
      <variable><type>int</type><name>drugid</name></variable>
      <variable><type>int</type><name>type</name></variable>
      <variable><type>double</type><name>myx</name></variable>
      <variable><type>double</type><name>myy</name></variable>
    </variables>
  </message>

  <message>
    <name>im_here</name>
    <variables>
      <variable><type>int</type><name>cellid</name></variable>
      <variable><type>int</type><name>mcat</name></variable>
      <variable><type>double</type><name>myx</name></variable>
      <variable><type>double</type><name>myy</name></variable>
    </variables>
  </message>
```

```
  <message>
    <name>cell_tot_msg</name>
    <variables>
      <variable><type>int</type><name>total_cells</name></variable>
    </variables>
  </message>
</messages>
</xmodel>

#include "header.h"
#include "Cell_agent_header.h"
#include "library_header.h"

/* add a state change
1) normal cells grow set state change quite low for A B any cell
can mutate to a,b,normal, a+b
*/

/* Cell functions */

int cell_here()
{
  add_im_here_message(MYID, MCAT, XPOS, YPOS);
  return 0;
}


int cell_divide()
{
  double cells_total=0;
  cell_tot_msg_message=get_first_cell_tot_msg_message();
  while(cell_tot_msg_message)
  {
    cells_total=cell_tot_msg_message->total_cells;
    cell_tot_msg_message = get_next_cell_tot_msg_message(cell_tot_msg_message);
  }
  double temp=0.0;
  double temp2=0.0;

  double c =(double)rand()/(double)RAND_MAX*5;
  double d=(double)rand()/(double)RAND_MAX*5;

  temp=random_double(0.0,1.0);
  double e =(double)rand()/(double)RAND_MAX*100;

  double current_population=0.0, pop_max=0.0;
  pop_max=2500;

  current_population=cells_total;
  double fraction=0.0;
  fraction=current_population/pop_max;
  double fraction2=0.0;
  fraction2=(1-fraction)*cell_division_rate;

  if(temp<fraction2)
  {
```

```
    if(e<25)
    {
      add_Cell_agent(1000,MCAT,XPOS+c,YPOS+d, ZPOS,10);
}
    else if(e<50)
    {
      add_Cell_agent(1000,MCAT,XPOS+c,YPOS-d, ZPOS,10);
    }
    else if(e<75)
    {
      add_Cell_agent(1000,MCAT,XPOS-c,YPOS+d, ZPOS,10);
}
else
{
  add_Cell_agent(1000,MCAT,XPOS-c,YPOS-d, ZPOS,10);
}
    }
  temp=random_double(0.0,1.0);
  temp2=random_double(0.0,1.0);
  if(temp<0.01)
  {
    if(MCAT==3)
    {
      if(temp2<0.01)
      {
        MCAT=3;
}
  else if(temp2<0.51)
      {
        MCAT=2;
      }
      else
      {
        MCAT=1;
      }
    }
    if((MCAT==1)||(MCAT==2))
    {
      if(temp2<0.01)
      {
        MCAT=0;
        printf("ZERO is PRODUCED!");
      }
    }
  }
  return 0;
}

int cell_check_drugs()
{
  int closest_drug_a_id = -1;
  int closest_drug_b_id = -1;
  double shortest_distance = 9999.0;
  double current_distance_squared;

  double drugx, drugy;
  double theta;
```

```
drug_a_information_message=get_first_drug_a_information_message();
while(drug_a_information_message)
{
  drugx=drug_a_information_message->myx;
  drugy=drug_a_information_message->myy;
  current_distance_squared = (drugx - XPOS)*(drugx - XPOS) +
      (drugy - YPOS)*(drugy - YPOS);
  if(current_distance_squared <=  (attack_length*attack_length))
  {
    if(current_distance_squared < shortest_distance)
    {
      shortest_distance = current_distance_squared;
      closest_drug_a_id = drug_a_information_message->drugid;
    }
  }
  drug_a_information_message = get_next_drug_a_information_message
                              (drug_a_information_message);
}

if(closest_drug_a_id != -1)
{
  if(shortest_distance <=  (attack_length*attack_length))
  {
if(MCAT==1)
    {
      add_drug_used_message(closest_drug_a_id, MCAT, XPOS, YPOS);
      return 1;
    }
    if(MCAT==3)
    {
      add_drug_used_message(closest_drug_a_id, MCAT, XPOS, YPOS);
      return 1;
    }
  }
}
drug_b_information_message=get_first_drug_b_information_message();
while(drug_b_information_message)
{
  drugx=drug_b_information_message->myx;
  drugy=drug_b_information_message->myy;
current_distance_squared = (drugx - XPOS)*(drugx - XPOS) +
          (drugy - YPOS)*(drugy - YPOS);
  if(current_distance_squared <=  (attack_length*attack_length))
  {
    if(current_distance_squared < shortest_distance)
    {
      shortest_distance = current_distance_squared;
      closest_drug_b_id = drug_b_information_message->drugid;
    }
  }
  drug_b_information_message = get_next_drug_b_information_message
                                (drug_b_information_message);
}
if(closest_drug_b_id != -1)
{
  if(shortest_distance <=  (attack_length*attack_length))
  {
```

```
      if(MCAT==2)
      {
        add_drug_used_message(closest_drug_b_id, MCAT, XPOS, YPOS);
        return 1;
      }
      if(MCAT==3)
      {
        add_drug_used_message(closest_drug_b_id, MCAT, XPOS, YPOS);
        return 1;
      }
    }
  }
  return 0;
}

int cell_decay()
{
  CLIFE=CLIFE-(CLIFE*cell_decay_prob);
  if(CLIFE<=0.1)
  {
    return 1;
  }
  return 0;
}


#include "header.h"
#include "DrugA_agent_header.h"
#include "library_header.h"


int drug_a_location()
{
  add_drug_a_information_message(MYID, XPOS, YPOS);
  return 0;
}

int drug_a_decay()
{
  ALIFE=ALIFE-(ALIFE*drug_a_decay_prob);
  if(ALIFE<0.1)
  {
    return 1;
  }
  return 0;
}

#include "header.h"
#include "DrugB_agent_header.h"
#include "library_header.h"

int drug_b_location()
{
  add_drug_b_information_message(MYID, XPOS, YPOS);
  return 0;
}
```

```
int drug_b_decay()
{
  BLIFE=BLIFE-(BLIFE*drug_b_decay_prob);
  if(BLIFE<0.1)
  {
    return 1;
  }
  return 0;
}
```

7.6.1   Using Multiple Outputs

The same model can be tested with multiple conditions to vary the drug introduction time. Figure 7.14 shows the plots of the normal cells versus drug-affected cells. The model results showed a competition between the cancer cells for survival. Figure 7.15 shows 2D and 3D views of the model results. It is important to be able to manipulate the results with multiple output tools to analyze them closely. Certain models need 3D views to show effect of surrounding cells, whereas some models can use graphs to show concentration gradients between cells and their survival rates. Multiple output codes can be run separate to simulations or during simulations as seen earlier in Drosophila modeling. Separate code can also be written, which will read or parse all the the XML simulation files to produce an excel data sheet. This can then easily extract the values into columns and draw graphs for analysis. FLAME has example files written in C and Java which can be run separately, after the simulation is finished to extract the values.

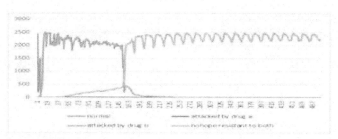

(a) Drug A introduced every 5 time steps and drug B every 15 time steps.

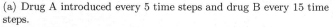

(b) Drug A and B introduced together every 10 time steps.

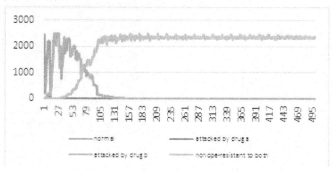

(c) Drug A introduced every 15 time steps and drug B every 5 time steps.

FIGURE 7.14: Testing drug effect on cancer cells.

(a) Snapshot of cancer cells at time 10.

(b) Snapshot of cancer cells at time 50.

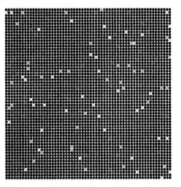

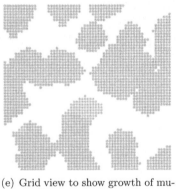

(c) Snapshot of cancer cells at time 500.

(d) Grid view of cells and drugs.

(e) Grid view to show growth of mutated cells.

FIGURE 7.15: Multiple views of the same cancer model.

Chapter 8

Testing Agent Behavior

8.1   Unit and System Testing 237
8.2   Statistical Testing of Data 239
8.3   Statistics Testing on Code 243
8.4   Testing Simulation Durations 244

While modeling is a complex task, testing of the models is a cumbersome task as well. The data being emergent results from many tiny interactions at lower scales, producing large changes on upper levels. This presents new challenges to find data anomalies or what causes upper level data to deviate from that expected. The questions posed are how and why system behaves in unpredictable ways.

Testing systems often involve reworking the model description, to find whether the model written was error-prone to begin with. It also involves analyzing large amounts of data produced as a result of running the simulations. Patterns can be studied to test and understand the behavior of the model, to see if rules were followed. Gilbert and Terna [73] described the use of verbal questioning to find general inconsistencies between various concepts and relationships.

8.1 Unit and System Testing

Agile methodologies teach that testing is not an end process but done through the development process. Various automated testing tools have been produced, suited for software being developed. Good coding practices make sure that the code is always testable at any stage of development.

- Automatically build the simulation code.

- Deploy the models to run.

- Test the data produced.

Test designs can help identify milestones, where domain experts can come together to check whether the model is being developed correctly.

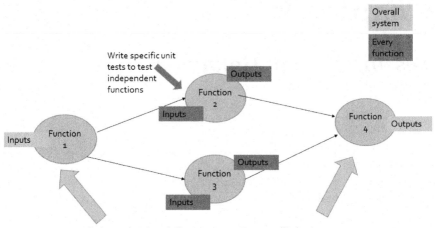

FIGURE 8.1: Testing low and high level functions.

Following test-driven design approaches, complete tests of the model can include unit tests along with integration testing of the whole system. Individual agent functions can be tested as unit tests, whereas the overall behavior can be tested using black box testing methods. Any system dependencies, such as agent messages, would come under system tests and not unit tests.

Unit testing of individual agent function can help isolate part of the agents and test single behavior (shown in Figure 8.1). It can show any immediate reasons for failure if they do not behave as expected. These should run quickly and provide the functional verification of the program. These agent functions can be tested independently with no dependency such as communicating with other agents. Trial messages can be simulated to test the function if needed. This helps signal any error at lowest levels of agent functions before running the whole model.

Testing the complete model as a black box is a much more complex activity. These tests start with some known preconditions and expected outputs of the system. Immediate data logs produced can be plotted in multiple forms, graphs or videos, and then analyzed to see if any wrong behavior is observed. This process would depend on simulation results rather than analyzing the code.

8.2 Statistical Testing of Data

Statistical testing procedures give invariants such as tools like DAIKON [58], which help identify the upper and lower bonds of the resulting agent variables through simulations. Adra et al. [1] used multi-objective optimization to generate sample starting conditions for models to simulate. The iterations were then sifted through, using DAIKON, to find the patterns of each memory variable. The rules produced were able to give maximum and minimum values for each variable, helping to find if any memory went out of allowable bounds. This showed that there were some discrepancies in the model code.

Results can be checked by studying the outputs produced such as graphs to help group large datasets into single pictures to find outlying data. Various evolutionary testing methods can generate test sets to test the outputs. This procedure involves using a computer program which uses genetic algorithms to vary the test code, changing the bounds according to the data produced. The resulting test case is the general test which can then be used to test the system.

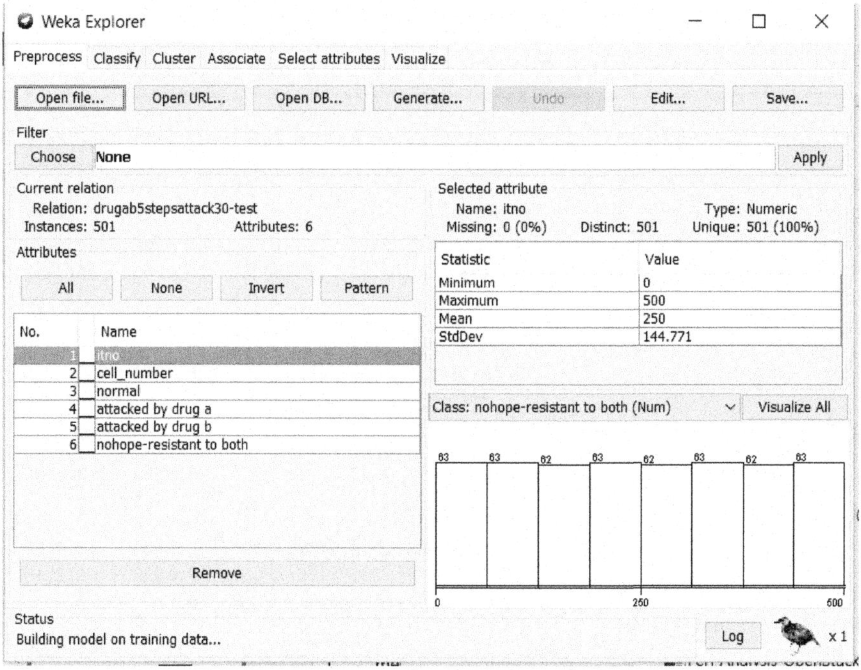

FIGURE 8.2: Screenshot of Weka analyzing cancer output.

Figure 8.2 shows the output of exporting the Excel data of the cancer model to study data patterns. The tool allows precomputed files in formats such as csv, xls or aiff to be loaded into its graphical output. Following is an example of running the clustering on the data to find overall statistical behaviors,

```
By running the clustering
== Run information ===
Scheme:weka.clusterers.EM -I 100 -N -1 -M 1.0E-6 -S 100
Relation:      drugab5stepsattack30-test
Instances:     501
Attributes:    6
               itno
               cell_number
               normal
               attacked by drug a
               attacked by drug b
               nohope-resistant to both
Test mode:evaluate on training data


=== Model and evaluation on training set ===


EM
==
Number of clusters selected by cross validation: 5
```

	Cluster				
Attribute	0	1	2	3	4
	(0.25)	(0.11)	(0.15)	(0.31)	(0.18)
itno					
mean	192.1514	82.0786	50.9843	421.6291	298.8974
std. dev.	36.4477	46.2217	21.0668	45.5711	25.4725
cell_number					
mean	2345.8669	1952.7397	2236.1558	2341.6396	2338.4172
std. dev.	90.3483	728.4002	179.8636	70.2953	75.7942
normal					
mean	0	227.4793	0	0	0
std. dev.	238.7914	673.3696	238.7914	238.7914	238.7914
attacked by drug a					
mean	0	2.8719	0	0	0
std. dev.	3.8596	11.0962	3.8596	3.8596	3.8596
attacked by drug b					
mean	107.4377	853.4992	2173.1369	15.5104	43.8573
std. dev.	28.541	584.4466	186.8803	9.1399	4.4213
nohope-resistant to both					
mean	2238.4292	868.8892	63.0189	2326.1293	2294.5599
std. dev.	87.904	663.6285	80.4388	70.191	74.7558

```
Time taken to build model (full training data) : 22.39 seconds

=== Model and evaluation on training set ===
```

```
Clustered Instances

0      125 ( 25%)
1       57 ( 11%)
2       73 ( 15%)
3      157 ( 31%)
4       89 ( 18%)

Log likelihood: -32.3838
```

The code was altered to add an anomaly in the code for allowing wrong values of Mcat to be generated.

```
//within the cell divide function
if(temp<0.01)
{
    if(MCAT==3)
    {
        if(temp2<0.01)
        {
            MCAT=3;
        }
        else if(temp2<0.51)
        {
            MCAT=2;
        }
        else if(temp2<0.98)
        {
            MCAT=1;
        }
        else //wrong conditions added
        {
            MCAT=4;
        }
    }
}
```

The simulation was executed again with the wrong code to see if this is detected from the data collected. The output is as follows:

```
=== Run information ===
Scheme:weka.clusterers.EM -I 100 -N -1 -M 1.0E-6 -S 100
Relation:     superimposed2results
Instances:    501
Attributes:   6
              itno
              cell_number
              normal
              attacked by drug a
              attacked by drug b
              nohope-resistant to both
```

```
Test mode:evaluate on training data

=== Model and evaluation on training set ===
EM
==
Number of clusters selected by cross validation: 5
```

	Cluster				
Attribute	0	1	2	3	4
	(0.02)	(0.27)	(0.29)	(0.05)	(0.37)
itno					
mean	7.9139	243.8688	106.8236	18.0812	406.5556
std. dev.	3.2163	41.6689	45.2054	9.313	54.2469
cell_number					
mean	741.0967	2344.6184	2337.2981	2332.0886	2342.4794
std. dev.	571.6462	110.9634	139.8895	193.7086	71.0952
normal					
mean	185.0694	0	0	475.5265	0
std. dev.	484.6112	239.2861	0.0001	937.0862	239.2861
attacked by drug a					
mean	0.7637	0	0	4.4818	0
std. dev.	1.9997	2.5182	2.5182	10.4813	2.5182
attacked by drug b					
mean	239.6292	0.0382	13.5826	415.5753	19.509
std. dev.	197.3125	1.3171	32.9128	269.2916	12.4946
nohope-resistant to both					
mean	0	2.1379	0	0	2322.9703
std. dev.	1126.6877	68.9101	0.1567	1126.6877	71.0925

```
Time taken to build model (full training data) : 16.86 seconds

=== Model and evaluation on training set ===
Clustered Instances

0        8 (  2%)
2      285 ( 57%)
3       20 (  4%)
4      188 ( 38%)

Log likelihood: -20.56071
```

With differing values of standard deviations and likelihoods it is still difficult to spot wrong model conditions. However the graphs in Figure 8.3 shows the conflicting picture with a spike in mutated cells suddenly with time. The graph, in this case, is better at showing that the model has some wrong conditions leading to erratic behavior.

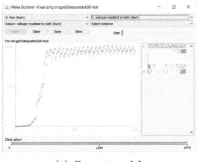

(a) Correct model.

(b) With wrong conditions added to code.

FIGURE 8.3: Plots of no-hope cells with correct and incorrect codes.

8.3 Statistics Testing on Code

The DAIKON inference tool is also found to be useful in finding code anomalies. The example from [106] shows the following code and invariants produced:

Code:

```
int ABS(int x)
{
  if (x>0) return x;
  else return (x*(-1));
}

int main ()
{
  int i=0;
  int abs_i;
  for (i=-5000;i<5000;i++)
  {
    abs_i=ABS(i);
  }
}
```

```
Expected invariants:
Return value of ABS(x) ==
(x>0) ? x: -x;
```

```
Daikon Output:
```

```
==================
std.ABS(int;):::ENTER
==================
std.ABS(int;):::EXIT1
x == return
==================
std.ABS(int;):::EXIT2
return == - x
==================
std.ABS(int;):::EXIT
x == orig(x)
x <= return
==================
```

The Daikon output being able to produce overall rules such as x will always be between certain values or how it relates to other variables in the code. Parsing through these files can again show how the variables will change during the simulations.

All of these testing methods do not deduce the causal factors for why 'bad' code sometimes goes undetected. The behavior in emergent systems are based on internal memory variables, and in some cases, variable bounds detected can help determine if certain variables behave wrongly in the simulation. For example, economic models, where wages always have to be above the minimum wage, can be parsed in multiple simulation results to see if this rule is violated by any one agent in the simulation. But these methods often carry the added complexity of extra code, processing time and difficulty in determining which part of the code is causing the values to behave in wrong manners.

8.4   Testing Simulation Durations

In addition to testing the outputs, from an HPC point of view, computer scientists are interested in studying the simulation time of these models to help find code bottlenecks and make it quicker to simulate. Figure 8.4 shows two models, the epithelium and sugarscape model running with multiple settings.

The epithelium model was run based on round-robin and geometric partitioning (Figure 8.4(a)) on different numbers of nodes. The results showed that round-robin partitioning performed better than geometric partitioning, which only performed well up to 16 nodes. Increasing the nodes to 32 resulted in the agents being too far away causing message communication overhead to increase and the simulation to slow down.

In Figure 8.4(b), the experiment involved 21,020 agents with 50 citizens, 1000 sugars and an Averager agent, which were multiplied by 20 scenes. The

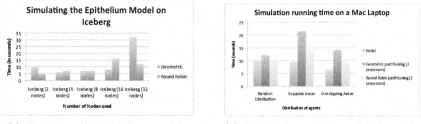

(a) Simulating the epithelium model.

(b) Basic sugarscape model by only changing the initial conditions.

FIGURE 8.4: Simulation times of models.

experiments were run in serial and in parallel on the Mac laptop machine. The parallel distribution was done in both geometric partitioning (partitions on geographic distribution using x and y positions of agents) and round-robin partitioning (partitions on agent numbers on available processors).

Figure 8.4(b) showed that even though the number of agents were the same, by changing initial distributions the simulation times changed. The random distribution took the least time because the agents, citizens and sugars could communicate locally on the same processors. However, when they were separated into different areas the agents had to communicate over different processors, thus causing an increase in simulation time. A similar result was seen in the overlapping areas being more than the random distribution times. These results showed that messaging between agents is a key factor for simulation times in agent-based modeling. How the agents are initially distributed influences their communications being either locally on the same processor or across nodes, which increases simulation time for messages to be sent across.

The graphs showed that simulation times can help identify certain bottlenecks but these vary depending on the model and the agents simulating.

Chapter 9

FLAME's Future

9.1   FLAME to FLAME GPU 247
    9.1.1   Visualizing Is Easy in FLAME GPU 273
    9.1.2   Utilizing Vector Calculations 276
9.2   Commercial Applications of FLAME 276

9.1   FLAME to FLAME GPU

The FLAME GPU version shares similar principles as FLAME, such as using an XML specification language as initial description of the models [162]. However, agent functions are written in C++, and interface with CUDA libraries so that they are processed on the local graphics card. The models showed an 80% improvement in simulation time and allowed real-time and 3D rendering of results while the simulations are running [108]. Each agent runs as an independent thread, performing its functions wrapped by the GPU kernel.

Although both versions of FLAME use similar methodologies and description languages, a number of changes were introduced in writing the models for every architecture. Most of the changes are due to basic capabilities of High Performance Computing (HPC) grid versus GPU cards. These changes are discussed, highlighting some issues raised by portability from modelers view:

Writing Agents. From the modeler perspective, implementing the system in FLAME in general was quite straight forward provided certain rules were being followed. Due to the synchronization nature of the framework, predefined before the simulation runs, FLAME does not allow agent behavior to contain loops, which may cause it to go back to a previous state. Similarly communications are decided at specific steps before the code is parallelized to ensure synchronization points (when the message boards are locked for reading and writing), to prevent any discordances in message data read, following a step-by-step progression in the model.

Pre-Allocation of Agent Memory. The grid FLAME version allowed memory to be allocated dynamically. This allowed complex agent memory such as using dynamic arrays or linked lists to be generated, as the model runs on the grid. This, however, is not the same on GPUs.

The GPU needs all memory allocated in advance, before the simulation starts, due to agents executed as threads on the cards. This means that all dynamic arrays in a model had to be changed to static arrays of specific sizes, causing considerable changes on the model, reducing some of the model complexity by limiting the memory size of agents.

In addition to the strict memory sizes, the grid version allows memory data structures of multiple data types to be created. This capability was not exhibited by FLAME GPU, allowing only a specific type of variable of fixed lengths data types. This raises issues if there are certain data structures used in the model which act as records of multiple datasets used by the agent such as a product inventory or biological rules. The cell model [35] was changed from a data structure into a 1D array and stored in memory where agents could globally read it serially. This caused a considerable rewriting of the model itself, however it did reduce the potential processing time of the model as data were being accessed serially rather than as dynamic data structures in the HPC version.

Message Communication between Agents. In FLAME HPC, signals can be passed between agents as messages with each agent creating multiple messages as needed. FLAME GPU limits this to only one message per function to be created, which limited the amount of information that was possible with one function. The model has to be broken down to allow multiple function transitions within one function to allow messages to be communicated, making the model quite complex or expanding the message at times to create much more information than previously designed. This constraint was introduced due to the manner in which the agent threads communicate, allowing only a single message to be synchronized per function.

Simpler versus Complex. Although both versions of grid and GPU are suitable for simulation, the GPU version seems more suitable for simpler model executions, with agents with basic memory allowing much faster simulations as compared to the grid version. This makes it suitable for game-based simulations to visualize the results quickly and where data analysis is not an extensive requirement. However, with more complex models such as the model of the complete European economy, with 15 different agent types with increasing complexity and multiple interactions, such models could not be processed on GPU cards.

Looping through Messages. Following is an example of looping through a message list as defined in FLAME HPC, which allows an agent to read through all messages in the list, and find agent id and state from the message and record it in memory.

```
int MyFunction (xmachine_memory* agent, xmachine_message_list* list)
{
```

```
    xmachine_message* message = get_first_message(list);
    while(message) {
       if (message->id == agent->id) {
          agent->state += message->state;
          return 0;
          }
       message = get_next_message(message, list);
       }
    return 0;
}
```

However the code does not work in FLAME GPU, causing the simulation to crash or hang. Instead, an extra flag 'finished' needs to be introduced to tell the code to leave the 'while' loop.

```
int MyFunction(xmahine_memory* agent, xmachine_message_list* list) {
    bool finished = false;
    xmachine_message* message = get_first_message(list);
    while(message) {
       if (!finished) {
          if (message->id == agent->id) {
             agent->state += message->state;
             finished = true;
          }
       }
       message = get_next_message(message, list);
    }
    return 0;
}
```

Discrete versus Continuous. In the GPU version, agents are defined as nature discrete or continuous. This affects how messages are parsed and handled during the simulation. This required modelers to understand this in advance, which was not seen in the grid version. The grid version allows easier writing of the agents and all are handled the same way.

Agent Birth and Death. Both architectures were able to handle agent addition similarly. Similar to the problem of dynamic memory allocation, agents can be introduced in the system if predetermined for the GPU. With every simulation the agent would be introduced using an environment agent, which keeps track of the maximum agents and generates a new agent by using the thread_id :$ID = Maxid + threadID$, where $threadID = blockIDx.x \times blockDimx + threadId.x$.

This guarantees that all generated IDs are unique (although not sequential), without using complicated atomic operations. The GPU, however, can only add one agent at a time step, whereas the Grid version could add multiple agents per step.

Real-time Visualization. The Grid version allows simulations to run as batch files producing results on disk. The results are later downloaded and processed to find data patterns in the simulations. The GPU was much simpler to do this, as it immediately integrates with a visualization engine. The agent could be observed as the simulation happens, and no time waits were needed to visualize the simulations.

Following is example code of an economic market in GPU:

```xml
<?xml version="1.0" encoding="utf-8"?>
<gpu:xmodel xmlns:gpu="http://www.dcs.shef.ac.uk/~paul/XMMLGPU"
  xmlns="http://www.dcs.shef.ac.uk/~paul/XMML">
  <name>Mariam Kiran</name>
  <gpu:environment>
  <gpu:functionFiles>
     <file>functions.c</file>
  </gpu:functionFiles>
  </gpu:environment>

  <xagents>
  <gpu:xagent>
  <name>Firm</name>
    <memory>
    <gpu:variable><type>int</type><name>id</name></gpu:variable>
    <gpu:variable><type>float</type><name>value</name></gpu:variable>
    <gpu:variable><type>float</type><name>a</name></gpu:variable>
    <gpu:variable><type>float</type><name>productivity</name></gpu:variable>
    <gpu:variable><type>float</type><name>profits</name></gpu:variable>
    <gpu:variable><type>float</type><name>f</name></gpu:variable>
    <gpu:variable><type>float</type><name>production</name></gpu:variable>
    <gpu:variable><type>int</type><name>goodsproduced</name></gpu:variable>
    <gpu:variable><type>int</type><name>stock</name></gpu:variable>
    <gpu:variable><type>int</type><name>sold</name></gpu:variable>
    <gpu:variable><type>int</type><name>labour</name></gpu:variable>
    <gpu:variable><type>int</type><name>numberofworkers</name></gpu:variable>
    <gpu:variable><type>int</type><name>workersize</name></gpu:variable>
    <gpu:variable><type>float</type><name>price</name></gpu:variable>
    <gpu:variable><type>float</type><name>oldprice</name></gpu:variable>
    <gpu:variable><type>float</type><name>priceinflation</name></gpu:variable>
    <gpu:variable><type>float</type><name>sprice</name></gpu:variable>
    <gpu:variable><type>float</type><name>lprice</name></gpu:variable>
    <gpu:variable><type>float</type><name>avewage</name></gpu:variable>
    <gpu:variable><type>float</type><name>totalwagebill</name></gpu:variable>

    <!--dynamic arrays are not allowed in GPU-->
    <!--<gpu:variable><type>int_array</type><name>mall_id</name></gpu:variable>-->
    <gpu:variable><type>int</type><name>mall_vacancy</name></gpu:variable>
    <gpu:variable><type>int</type><name>mall_goods</name></gpu:variable>
    <gpu:variable><type>float</type><name>range</name></gpu:variable>
    <gpu:variable><type>float</type><name>x</name></gpu:variable>
    <gpu:variable><type>float</type><name>y</name></gpu:variable>
    <gpu:variable><type>float</type><name>z</name></gpu:variable>
    </memory>

    <functions>
    <gpu:function>
      <name>Firm_1</name>
      <currentState>f_one</currentState>
```

```
    <nextState>f_one</nextState>
    <outputs>
      <gpu:output>
        <messageName>vacancy</messageName>
        <gpu:type>single_message</gpu:type>
      </gpu:output>
    </outputs>

    <gpu:reallocate>false</gpu:reallocate>
    <gpu:RNG>true</gpu:RNG>
</gpu:function>

<gpu:function>
  <name>Firm_1_b</name>
  <!--New function to send another message-->
  <currentState>f_one</currentState>
  <nextState>f_one</nextState>
  <outputs>
    <gpu:output>
      <messageName>priceinflation</messageName>
      <gpu:type>single_message</gpu:type>
    </gpu:output>
  </outputs>
  <gpu:reallocate>false</gpu:reallocate>
  <gpu:RNG>false</gpu:RNG>
</gpu:function>

<gpu:function>
  <name>Firm_3</name>
  <currentState>f_one</currentState>
  <nextState>f_one</nextState>
  <inputs>
    <gpu:input><messageName>employed</messageName></gpu:input>
  </inputs>
  <outputs>
    <gpu:output>
      <messageName>firm_stock_price</messageName>
      <gpu:type>optional_message</gpu:type>
    </gpu:output>
  </outputs>
  <gpu:reallocate>false</gpu:reallocate>
  <gpu:RNG>true</gpu:RNG>
</gpu:function>

<gpu:function>
  <name>Firm_4</name>
  <currentState>f_one</currentState>
  <nextState>f_one</nextState>
  <inputs>
    <gpu:input><messageName>firm_stock</messageName></gpu:input>
  </inputs>
  <gpu:reallocate>false</gpu:reallocate>
  <gpu:RNG>false</gpu:RNG>
</gpu:function>
</functions>

<states>
```

```
      <gpu:state><name>f_one</name></gpu:state>
      <initialState>f_one</initialState>
   </states>
   <gpu:type>continuous</gpu:type>
   <gpu:bufferSize>65535</gpu:bufferSize>
</gpu:xagent>

<gpu:xagent>
<name>Person</name>
   <memory>
   <gpu:variable><type>int</type><name>id</name></gpu:variable>
   <gpu:variable><type>float</type><name>savings</name></gpu:variable>
   <gpu:variable><type>float</type><name>wage</name></gpu:variable>
   <gpu:variable><type>int</type><name>firmid</name></gpu:variable>
   <gpu:variable><type>int</type><name>mall_application</name></gpu:variable>
   <gpu:variable><type>int</type><name>mall_shopping</name></gpu:variable>
   <gpu:variable><type>float</type><name>range</name></gpu:variable>
   <gpu:variable><type>float</type><name>x</name></gpu:variable>
   <gpu:variable><type>float</type><name>y</name></gpu:variable>
   <gpu:variable><type>float</type><name>z</name></gpu:variable>
   <memory>

   <functions>
   <gpu:function>
      <name>Person_1</name>
      <currentState>p_one</currentState>
      <nextState>p_one</nextState>
      <inputs>
        <gpu:input><messageName>priceinflation</messageName></gpu:input>
      </inputs>
      <outputs>
        <gpu:output>
          <messageName>application</messageName>
          <gpu:type>optional_message</gpu:type>
        </gpu:output>
      </outputs>
      <gpu:reallocate>false</gpu:reallocate>
      <gpu:RNG>true</gpu:RNG>
   </gpu:function>

   <gpu:function>
      <name>Person_2</name>
      <currentState>p_one</currentState>
      <nextState>p_one</nextState>
      <inputs>
        <gpu:input><messageName>employed</messageName></gpu:input>
      </inputs>
      <outputs>
        <gpu:output>
          <messageName>consumer_spending</messageName>
          <gpu:type>optional_message</gpu:type>
        </gpu:output>
      </outputs>
      <gpu:reallocate>false</gpu:reallocate>
      <gpu:RNG>true</gpu:RNG>
   </gpu:function>
```

```
<gpu:function>
  <name>Person_4</name>
  <currentState>p_one</currentState>
  <nextState>p_one</nextState>
  <inputs>
    <gpu:input><messageName>consumer_spent</messageName></gpu:input>
  </inputs>
  <gpu:reallocate>false</gpu:reallocate>
  <gpu:RNG>false</gpu:RNG>
</gpu:function>
</functions>

<states>
  <gpu:state><name>p_one</name></gpu:state>
  <initialState>p_one</initialState>
</states>
<gpu:type>continuous</gpu:type>
<gpu:bufferSize>65535</gpu:bufferSize>
</gpu:xagent>

<gpu:xagent>
<name>Mall</name>
  <memory>
  <gpu:variable><type>int</type><name>id</name></gpu:variable>

  <!-- Cannot have arrays so need individual data variables-->
  <gpu:variable><type>int</type><name>firm_1</name></gpu:variable>
  <gpu:variable><type>int</type><name>firm_1_vacancy</name></gpu:variable>
  <gpu:variable><type>int</type><name>firm_2</name></gpu:variable>
  <gpu:variable><type>int</type><name>firm_2_vacancy</name></gpu:variable>
  <gpu:variable><type>int</type><name>firm_3</name></gpu:variable>
  <gpu:variable><type>int</type><name>firm_3_vacancy</name></gpu:variable>
  <gpu:variable><type>int</type><name>firm_4</name></gpu:variable>
  <gpu:variable><type>int</type><name>firm_4_vacancy</name></gpu:variable>
  <gpu:variable><type>int</type><name>firm_5</name></gpu:variable>
  <gpu:variable><type>int</type><name>firm_5_vacancy</name></gpu:variable>
  <gpu:variable><type>int</type><name>firm_1_goods</name></gpu:variable>
  <gpu:variable><type>float</type><name>firm_1_price</name></gpu:variable>
  <gpu:variable><type>int</type><name>firm_1_stock</name></gpu:variable>
  <gpu:variable><type>int</type><name>firm_2_goods</name></gpu:variable>
  <gpu:variable><type>float</type><name>firm_2_price</name></gpu:variable>
  <gpu:variable><type>int</type><name>firm_2_stock</name></gpu:variable>
  <gpu:variable><type>int</type><name>firm_3_goods</name></gpu:variable>
  <gpu:variable><type>float</type><name>firm_3_price</name></gpu:variable>
  <gpu:variable><type>int</type><name>firm_3_stock</name></gpu:variable>
  <gpu:variable><type>int</type><name>firm_4_goods</name></gpu:variable>
  <gpu:variable><type>float</type><name>firm_4_price</name></gpu:variable>
  <gpu:variable><type>int</type><name>firm_4_stock</name></gpu:variable>
  <gpu:variable><type>int</type><name>firm_5_goods</name></gpu:variable>
  <gpu:variable><type>float</type><name>firm_5_price</name></gpu:variable>
  <gpu:variable><type>int</type><name>firm_5_stock</name></gpu:variable>
  <gpu:variable><type>int</type><name>total_vacancies</name></gpu:variable>
  <gpu:variable><type>float</type><name>range</name></gpu:variable>
  <gpu:variable><type>float</type><name>x</name></gpu:variable>
  <gpu:variable><type>float</type><name>y</name></gpu:variable>
  <gpu:variable><type>float</type><name>z</name></gpu:variable>
  </memory>
```

```
<functions>
<gpu:function>
  <name>Spread_awareness</name>
  <currentState>m_one</currentState>
  <nextState>m_one</nextState>
  <gpu:reallocate>false</gpu:reallocate>
  <gpu:RNG>false</gpu:RNG>
</gpu:function>

<gpu:function>
  <name>Job_market</name>
  <currentState>m_one</currentState>
  <nextState>m_one</nextState>
  <inputs>
    <gpu:input><messageName>vacancy</messageName></gpu:input>
  </inputs>
  <gpu:reallocate>false</gpu:reallocate>
  <gpu:RNG>false</gpu:RNG>
</gpu:function>

<gpu:function>
  <name>Job_market_b</name>
  <currentState>m_one</currentState>
  <nextState>m_one</nextState>
  <inputs>
    <gpu:input><messageName>application</messageName></gpu:input>
  </inputs>
  <outputs>
    <gpu:output>
      <messageName>employed</messageName>
      <gpu:type>optional_message</gpu:type>
    </gpu:output>
  </outputs>
  <gpu:reallocate>false</gpu:reallocate>
  <gpu:RNG>false</gpu:RNG>
</gpu:function>

<gpu:function>
  <name>Goods_market</name>
  <currentState>m_one</currentState>
  <nextState>m_one</nextState>
  <inputs>
    <gpu:input><messageName>firm_stock_price</messageName></gpu:input>
  </inputs>
  <gpu:reallocate>false</gpu:reallocate>
  <gpu:RNG>false</gpu:RNG>
</gpu:function>

<gpu:function>
  <name>Goods_market_b</name>
  <currentState>m_one</currentState>
  <nextState>m_one</nextState>
  <inputs>
    <gpu:input><messageName>consumer_spending</messageName></gpu:input>
  </inputs>
  <outputs>
```

```
      <gpu:output>
        <messageName>consumer_spent</messageName>
        <gpu:type>optional_message</gpu:type>
      </gpu:output>
    </outputs>
    <gpu:reallocate>false</gpu:reallocate>
    <gpu:RNG>false</gpu:RNG>
  </gpu:function>

  <gpu:function>
    <name>Goods_market_c</name>
    <currentState>m_one</currentState>
    <nextState>m_one</nextState>
    <outputs>
      <gpu:output>
        <messageName>firm_stock</messageName>
        <gpu:type>optional_message</gpu:type>
      </gpu:output>
    </outputs>
    <gpu:reallocate>false</gpu:reallocate>
    <gpu:RNG>false</gpu:RNG>
  </gpu:function>

  </functions>

  <states>
    <gpu:state><name>m_one</name></gpu:state>
    <initialState>m_one</initialState>
  </states>
  <gpu:type>continuous</gpu:type>
  <gpu:bufferSize>65535</gpu:bufferSize>
</gpu:xagent>

</xagents>

<messages>
  <gpu:message>
  <name>mall_location</name>
  <description>Mall location message</description>
  <variables>
    <gpu:variable><type>int</type><name>mall_id</name></gpu:variable>
    <gpu:variable><type>float</type><name>range</name></gpu:variable>
    <gpu:variable><type>float</type><name>x</name></gpu:variable>
    <gpu:variable><type>float</type><name>y</name></gpu:variable>
    <gpu:variable><type>float</type><name>z</name></gpu:variable>
  </variables>
  <gpu:partitioningNone/>
  <gpu:bufferSize>65536</gpu:bufferSize>
  </gpu:message>

  <gpu:message>
  <name>priceinflation</name>
  <description>Posted by firm when it calculates the next price of goods.
  The message is read by the workers to help calculate their new wages,
  because they consider the price inflation to do this</description>
  <variables>
    <gpu:variable><type>int</type><name>firm_id</name></gpu:variable>
```

```
  <gpu:variable><type>float</type><name>priceinflation</name></gpu:variable>
  <gpu:variable><type>float</type><name>range</name></gpu:variable>
  <gpu:variable><type>float</type><name>x</name></gpu:variable>
  <gpu:variable><type>float</type><name>y</name></gpu:variable>
  <gpu:variable><type>float</type><name>z</name></gpu:variable>
</variables>
<gpu:partitioningNone/>
<gpu:bufferSize>65536</gpu:bufferSize>
</gpu:message>

<!--    Message for applying to firm   -->
<gpu:message>
<name>application</name>
<description>This message is posted by the worker, applying to firm,
with intended wage</description>
<variables>
  <gpu:variable><type>int</type><name>person_id</name></gpu:variable>
  <gpu:variable><type>float</type><name>person_wage</name></gpu:variable>
  <gpu:variable><type>int</type><name>mall_id</name></gpu:variable>
  <gpu:variable><type>float</type><name>range</name></gpu:variable>
  <gpu:variable><type>float</type><name>x</name></gpu:variable>
  <gpu:variable><type>float</type><name>y</name></gpu:variable>
  <gpu:variable><type>float</type><name>z</name></gpu:variable>
</variables>
<gpu:partitioningNone/>
<gpu:bufferSize>65536</gpu:bufferSize>
</gpu:message>

<gpu:message>
<name>vacancy</name>
<description>Message for firm vacancies</description>
<variables>
  <gpu:variable><type>int</type><name>firm_id</name></gpu:variable>
  <gpu:variable><type>int</type><name>vacancies</name></gpu:variable>
  <gpu:variable><type>int</type><name>mall_id</name></gpu:variable>
  <gpu:variable><type>float</type><name>range</name></gpu:variable>
  <gpu:variable><type>float</type><name>x</name></gpu:variable>
  <gpu:variable><type>float</type><name>y</name></gpu:variable>
  <gpu:variable><type>float</type><name>z</name></gpu:variable>
</variables>
<gpu:partitioningNone/>
<gpu:bufferSize>65536</gpu:bufferSize>
</gpu:message>

<gpu:message>
<name>employed</name>
<description>Sent by firms to let the workers know who are
employed and by whom</description>
<variables>
  <gpu:variable><type>int</type><name>person_id</name></gpu:variable>
  <gpu:variable><type>float</type><name>person_wage</name></gpu:variable>
  <gpu:variable><type>int</type><name>firm_id</name></gpu:variable>
  <gpu:variable><type>float</type><name>range</name></gpu:variable>
  <gpu:variable><type>float</type><name>x</name></gpu:variable>
  <gpu:variable><type>float</type><name>y</name></gpu:variable>
  <gpu:variable><type>float</type><name>z</name></gpu:variable>
</variables>
```

```
<gpu:partitioningNone/>
<gpu:bufferSize>65536</gpu:bufferSize>
</gpu:message>

<gpu:message>
<name>consumer_spending</name>
<description>From Mall indicating how much to spend</description>
<variables>
  <gpu:variable><type>int</type><name>person_id</name></gpu:variable>
  <gpu:variable><type>float</type><name>spending</name></gpu:variable>
  <gpu:variable><type>int</type><name>mall_id</name></gpu:variable>
  <gpu:variable><type>float</type><name>range</name></gpu:variable>
  <gpu:variable><type>float</type><name>x</name></gpu:variable>
  <gpu:variable><type>float</type><name>y</name></gpu:variable>
  <gpu:variable><type>float</type><name>z</name></gpu:variable>
</variables>
<gpu:partitioningNone/>
<gpu:bufferSize>65536</gpu:bufferSize>
</gpu:message>

<gpu:message>
<name>consumer_spent</name>
<description>From Mall indicating how much has been spent</description>
<variables>
  <gpu:variable><type>int</type><name>person_id</name></gpu:variable>
  <gpu:variable><type>float</type><name>spent</name></gpu:variable>
  <gpu:variable><type>float</type><name>range</name></gpu:variable>
  <gpu:variable><type>float</type><name>x</name></gpu:variable>
  <gpu:variable><type>float</type><name>y</name></gpu:variable>
  <gpu:variable><type>float</type><name>z</name></gpu:variable>
</variables>
<gpu:partitioningNone/>
<gpu:bufferSize>65536</gpu:bufferSize>
</gpu:message>

<gpu:message>
<name>firm_stock</name>
<description>Let people know how much stock is left at firm</description>
<variables>
  <gpu:variable><type>int</type><name>firm_id</name></gpu:variable>
  <gpu:variable><type>int</type><name>stock</name></gpu:variable>
  <gpu:variable><type>float</type><name>range</name></gpu:variable>
  <gpu:variable><type>float</type><name>x</name></gpu:variable>
  <gpu:variable><type>float</type><name>y</name></gpu:variable>
  <gpu:variable><type>float</type><name>z</name></gpu:variable>
</variables>
<gpu:partitioningNone/>
<gpu:bufferSize>65536</gpu:bufferSize>
</gpu:message>

<gpu:message>
<name>firm_stock_price</name>
<description>Let people know the price of the left stock</description>
<variables>
  <gpu:variable><type>int</type><name>firm_id</name></gpu:variable>
  <gpu:variable><type>int</type><name>stock</name></gpu:variable>
  <gpu:variable><type>float</type><name>price</name></gpu:variable>
```

```
      <gpu:variable><type>int</type><name>mall_id</name></gpu:variable>
      <gpu:variable><type>float</type><name>range</name></gpu:variable>
      <gpu:variable><type>float</type><name>x</name></gpu:variable>
      <gpu:variable><type>float</type><name>y</name></gpu:variable>
      <gpu:variable><type>float</type><name>z</name></gpu:variable>
    </variables>
    <gpu:partitioningNone/>
    <gpu:bufferSize>65536</gpu:bufferSize>
    </gpu:message>
  </messages>

  <!-- Order of function execution-->
  <layers>
    <layer>
      <gpu:layerFunction><name>Spread_awareness</name></gpu:layerFunction>
    </layer>
    <layer>
      <gpu:layerFunction><name>Firm_1</name></gpu:layerFunction>
    </layer>
    <layer>
      <gpu:layerFunction><name>Firm_1_b</name></gpu:layerFunction>
    </layer>
    <layer>
      <gpu:layerFunction><name>Person_1</name></gpu:layerFunction>
    </layer>
    <layer>
      <gpu:layerFunction><name>Job_market</name></gpu:layerFunction>
    </layer>
    <layer>
      <gpu:layerFunction><name>Job_market_b</name></gpu:layerFunction>
    </layer>
    <layer>
      <gpu:layerFunction><name>Firm_3</name></gpu:layerFunction>
    </layer>
    <layer>
      <gpu:layerFunction><name>Person_2</name></gpu:layerFunction>
    </layer>
    <layer>
      <gpu:layerFunction><name>Goods_market</name></gpu:layerFunction>
    </layer>
    <layer>
      <gpu:layerFunction><name>Goods_market_b</name></gpu:layerFunction>
    </layer>
    <layer>
      <gpu:layerFunction><name>Goods_market_c</name></gpu:layerFunction>
    </layer>
    <layer>
      <gpu:layerFunction><name>Firm_4</name></gpu:layerFunction>
    </layer>
    <layer>
      <gpu:layerFunction><name>Person_4</name></gpu:layerFunction>
    </layer>
  </layers>
</gpu:xmodel>

/**** the functions C file***/
#ifndef _FUNCTIONS_H_
```

```
#define _FUNCTIONS_H_
#include "header.h"

/** \def no_firms_to_apply_to
 * \brief The number of firms for workers to apply to. */
#define NO_FIRMS_TO_APPLY_TO 10

/** \def no_firms_to_buy_from
 * \brief The number of firms for workers to buy goods from. */
#define no_firms_to_buy_from 10
#define printlog 0
#define CHANGEINPRODUCTIVITY 0.0
#define person_firm_distance 50.0
#define firm_mall_distance 50.0
#define person_mall_distance 50.0
#define MESSAGE_RANGE 50.0
#define MALLSNO 3
#define FIRMSNO 5
#define PEOPLENO 30
#define WORKERSNO 30

//Random number generator
inline __device__ float GetRandomNumber(const float minNumber,
                    const float maxNumber, RNG_rand48* rand48)
{
  return minNumber+(rnd(rand48)*(maxNumber-minNumber));
}

/** \fn void bubble_sort(int * id, double * wage, int length)
 * \brief Use bubble sorting algorithm to sort worker list by wage
 * \param id Pointer to the list of worker ids.
 * \param wage Pointer to the associated list of worker wages.
 * \param length Length of the lists.
 */
inline __device__ void bubble_sort(int * id, float * wage, int length)
{
  int i, j, tmpid;
  float tmpwage;

  /* Using bubble sorts nested loops */
  for(i=0; i<(length-1); i++)
  {
    for(j=0; j<(length-1)-i; j++)
    {
      /* Comparing the wages between neighbours */
      if(*(wage+j+1) < *(wage+j))
      {
        /* Check for end of list */
        if(*(wage+j+1) != 0)
        {
          /* Swap worker id and wage to move
             cheaper workers to the top of the list */
          tmpwage = *(wage+j);
          tmpid = *(id+j);
          *(wage+j) = *(wage+j+1);
          *(id+j) = *(id+j+1);
          *(wage+j+1) = tmpwage;
```

```
          *(id+j+1) = tmpid;
        }
      }
    }
  }
}


/** \fn void bubble_sort(int * id, double * wage, int length)
 * \brief Use bubble sorting algorithm to sort worker by wage
 * \param id Pointer to the list of worker ids.
 * \param wage Pointer to the associated list of worker wages.
 * \param length Length of the lists.
 */
void sort_prices(int * id, int * stock, float * price, int length)
{
  int i, j, tmpid, tmpstock;
  float tmpprice;

  /* Using bubble sorts nested loops */
  for(i=0; i<(length-1); i++)
  {
    for(j=0; j<(length-1)-i; j++)
    {
      /* Comparing the wages between neighbours */
      if(*(price+j+1) < *(price+j))
      {
        /* Check for end of list */
        if(*(price+j+1) != 0)
        {
          /* Swap worker id and wage to move
          cheaper workers to the top of the list */
          tmpprice = *(price+j);
          tmpid = *(id+j);
          tmpstock = *(stock+j);
          *(price+j) = *(price+j+1);
          *(id+j) = *(id+j+1);
          *(stock+j) = *(stock+j+1);
          *(price+j+1) = tmpprice;
          *(id+j+1) = tmpid;
          *(stock+j+1) = tmpstock;
        }
      }
    }
  }
}


/** \fn Firm_1()
 * \brief Set initial values of financial viability, production
 *\ and number of labour required during every iteration
 */
__FLAME_GPU_FUNC__ int Firm_1(xmachine_memory_Firm* agent,
        xmachine_message_vacancy_list* vacancy_messages,RNG_rand48* rand48)
{
  float mean=0.0f;
  float random_no=rnd(rand48);
  int mall_chosen;
```

```
agent->a=agent->a + ((1.0 - agent->f)* agent->profits);
mean = agent->f /(agent->price * agent->productivity);
agent->productivity=agent->productivity + CHANGEINPRODUCTIVITY;

if(random_no<0.5)
{
  if((agent->stock == 0) && (agent->goodsproduced > 0))
  {
    agent->sprice=agent->price *(1.0 + random_no);
  }
  if((agent->stock > 0) || (agent->goodsproduced == 0))
  {
    agent->sprice=agent->price *(1.0 - random_no);
  }
}
else
{
  if((agent->stock == 0) && (agent->goodsproduced > 0))
  {
    agent->production=agent->production * (1.0 + random_no);
  }
  if(agent->stock > 0)
  {
    agent->production=agent->production * (1.0 - random_no);
  }
}
agent->labour= (int)(agent->production/agent->productivity);

if(agent->labour < 1)
{
  agent->labour=1;
}

/* Apply to a random Mall */
mall_chosen=floor(GetRandomNumber(1,MALLSNO+1,rand48));
if(mall_chosen>0)
{
  agent->mall_vacancy=mall_chosen;
  add_vacancy_message(vacancy_messages,agent->id, agent->labour,
      agent->mall_vacancy, MESSAGE_RANGE, agent->x, agent->y, 0.0f);
}
else
{
  agent->mall_vacancy=0;
}
return 0;
}


__FLAME_GPU_FUNC__ int Firm_1_b(xmachine_memory_Firm* agent,
          xmachine_message_priceinflation_list* priceinflation_messages)
{
  /* Post message of the price inflation (calculated as a
  // difference between the previous and present price)
  This is later used to calculate the wages of the workers */

  add_priceinflation_message(priceinflation_messages, agent->id,
      agent->priceinflation, MESSAGE_RANGE, agent->x, agent->y, 0.0f);
```

```
  return 0;
}

/** \fn Firm_3()
 * \brief Calculates the total wage bill of the workers,
 * lowest price of the goods, and the amount of stock
 */
__FLAME_GPU_FUNC__ int Firm_3(xmachine_memory_Firm* agent,
             xmachine_message_employed_list* employed_messages,
             xmachine_message_firm_stock_price_list* firm_stock_price_messages,
             RNG_rand48* rand48)
{
  float wage_bill = 0.0f;
  float lowest_price = 0.0f;
  agent->workersize=0;
  /* Reset the hired worker lists to zero */
  //xmemory->workerid.size = 0;
  //xmemory->workerwage.size = 0;
  float random_no=rnd(rand48);
  int mall_chosen;
  //int workeridarray[20];
  //float workerwagearray[20];

  /* Read employment messages */
  xmachine_message_employed* current_message = get_first_employed_message
                                          (employed_messages);
  while(current_message)
  {
    /* If the employment message refers to this worker */
    if(agent->id == current_message->firm_id)
    {
      /* Update workers firm id */
      //save upto 20 workers
      if(agent->workersize<20)
      {
        agent->workersize=agent->workersize++;
        wage_bill += current_message->person_wage;
      }
    }
    /* Move onto next message to check */
    current_message
      = get_next_employed_message(current_message,employed_messages);
  }
  agent->numberofworkers=agent->workersize;
  agent->avewage=wage_bill / agent->numberofworkers;
  agent->totalwagebill=wage_bill;

  /* calculate lowest price for goods - Refer to equation (5) */
  lowest_price = wage_bill / (int)(agent->productivity);
  /* save the old price of the goods before changing them */
  agent->oldprice=agent->price;
  /* Calculate new offer price - if the satisticing price is less than
  the lowest price then set the price to the lowest price the firm can have.
  This lowest price is the lowest the firm can keep to stop suffering losses
  else use the satisficing price to gain profits
  (Rule stated in the paper to prevent firm from suffering loss) */
  if(agent->sprice < lowest_price)
```

```
      {
        agent->price=lowest_price;
      }
      else
      {
        agent->price=agent->sprice;
      }

      /* Calculate price inflation by calculating the difference of
      the prices from the new set price and the old */
      agent->priceinflation=agent->price - agent->oldprice;

      /* Send message stating offer price for goods with firm id
      and the offer price */
      /* Send message stating amount of stock to be sold by the firm
      goodsproduced is an integer value of the productivity factor
      multiplied by the man power the firm has - Refer to equation(2) */
      agent->goodsproduced=(int)(agent->productivity * agent->numberofworkers);

      /* Set the stock to be equal to the goods produced */
      agent->stock=agent->goodsproduced;
      /* Post message that firm is selling with firm id and the stock it has */
      /*add_firm_stock_message(get_id(), get_stock());*/

      /* New */
      /* Goods to a random Mall */
      mall_chosen=floor(GetRandomNumber(1,MALLSNO+1,rand48));
      if((mall_chosen>0)&&(agent->stock > 0))
      {
        add_firm_stock_price_message(firm_stock_price_messages,agent->id,
                  agent->stock, agent->price, agent->mall_goods, MESSAGE_RANGE,
                      agent->x, agent->y, 0.0f);
      }
      else
      {
        agent->mall_goods=0;
      }
      return 0;
}


/** \fn Firm_4()
 * \brief Calculates the stock being sold by the firm,
 the revenue and the profits its earned*/
__FLAME_GPU_FUNC__ int Firm_4(xmachine_memory_Firm* agent,
              xmachine_message_firm_stock_list* firm_stock_messages)
{
  int revenue;
  float ave_wage;

  /* Find last stock message for firm says how much stock left */
  xmachine_message_firm_stock* current_message =
              get_first_firm_stock_message(firm_stock_messages);
  while(current_message)
  {
    /* If the stock message relates to this firm */
    if(agent->id == current_message->firm_id)
    {
```

```
      /* Set the stock amount */
      agent->stock=current_message->stock;
      /* Exit stock message list */
      current_message = NULL;
    }
    /* Else carry on reading stock messages */
    else
      current_message = get_next_firm_stock_message
                  (current_message,firm_stock_messages);
  }


  /* Set the amount of stock sold */
  agent->sold=agent->goodsproduced - agent->stock;
  /* Calculate the revenue earned by the firm */
  revenue = (agent->goodsproduced - agent->stock) * agent->price;
  /* Calculate wage bill DONE EARLIER
  for(i=0;i<agent->numberofworkers;i++)
  {
    totalwagebill += xmemory->workerwage.array[i];
  }*/


  /* Calculate the profits */
  agent->profits=revenue-agent->totalwagebill;


  /* Calculate R&D fraction */
  /* Calculate net worth */
  /* Calculate the average wage, just for information */
  if(agent->numberofworkers == 0)
  {
    ave_wage = 0;
  }
  else
  {
    ave_wage = agent->totalwagebill/(float)agent->numberofworkers;
  }
  agent->avewage=ave_wage;
  /* Print results as simulation progresses */
  return 0;
}


/** \fn Person_1()
 * \brief Calculates new wage and sends job applications
 */
__FLAME_GPU_FUNC__ int Person_1(xmachine_memory_Person* agent,
      xmachine_message_priceinflation_list* priceinflation_messages,
      xmachine_message_application_list* application_messages,
      RNG_rand48* rand48)
{
  int number_of_firms;
  int m = NO_FIRMS_TO_APPLY_TO;
  float price_inflation = 0.0f;
  int mall_chosen=0;
  number_of_firms = 0;
  xmachine_message_priceinflation* current_message =
              get_first_priceinflation_message(priceinflation_messages);
  while(current_message)
  {
```

```
      price_inflation += current_message->priceinflation;
      number_of_firms++;
      current_message = get_next_priceinflation_message(current_message,
                                        priceinflation_messages);
  }
  price_inflation = price_inflation / number_of_firms;
  if(number_of_firms < m)
  {
    m = number_of_firms;
  }
  if(price_inflation < -0.05f) price_inflation = -0.05f;
  if(price_inflation > 0.05f) price_inflation = 0.05f;
  if(agent->firmid == 0)
  {
    agent->wage=agent->wage * ( 1.0 + price_inflation ) * ( 1.0 - 0.0);
  }
  else
  {
    agent->wage=agent->wage * ( 1.0 + price_inflation ) * ( 1.0 + 0.01);
  }
  /* Apply to a random Mall */
  mall_chosen=floor(GetRandomNumber(1,MALLSNO+1,rand48));
  if(mall_chosen > 0)
  {
    agent->mall_application=mall_chosen;
    add_application_message(application_messages,agent->id,
            agent->wage, agent->mall_application, MESSAGE_RANGE,
            agent->x,agent->y, 0.0f);
  }
  else
  {
    agent->mall_application=0;
  }
  return 0;
}

/** \fn Person_2()
 * \brief Update employment status*/
__FLAME_GPU_FUNC__ int Person_2(xmachine_memory_Person* agent,
          xmachine_message_employed_list* employed_messages,
          xmachine_message_consumer_spending_list* consumer_spending_messages,
          RNG_rand48* rand48)
{
  float random_no=rnd(rand48);
  int mall_chosen=floor(GetRandomNumber(1,MALLSNO+1,rand48));
  agent->firmid=0;
  xmachine_message_employed* current_message =
                    get_first_employed_message(employed_messages);
  while(current_message)
  {
    if(agent->id == current_message->person_id)
    {
      agent->firmid=current_message->firm_id;
    }

    current_message
      = get_next_employed_message(current_message,employed_messages);
```

```
  }

  if(mall_chosen>0)
  {
    agent->mall_shopping=mall_chosen;
    add_consumer_spending_message(consumer_spending_messages,
            agent->id, agent->savings, agent->mall_shopping,
            MESSAGE_RANGE, agent->x, agent->y, 0.0f);
  }
  else
  {
    agent->mall_application=0;
  }
  return 0;
}


/** \fn Person_4()
 * \brief Add wage to savings */
__FLAME_GPU_FUNC__ int Person_4(xmachine_memory_Person* agent,
            xmachine_message_consumer_spent_list* consumer_spent_messages)
{
  /* See how much was spent at the Mall shopping */
  /* Find last stock message for firm says how much stock left */
  xmachine_message_consumer_spent* current_message =
            get_first_consumer_spent_message(consumer_spent_messages);
  while(current_message)
  {
    /* If the stock message relates to this firm */
    if(agent->id == current_message->person_id)
    {
      /* Set the stock amount */
      agent->savings=current_message->spent;
      /* Exit stock message list */
      current_message = NULL;
    }
    /* Else carry on reading stock messages */
    else
    {
      current_message = get_next_consumer_spent_message
                      (current_message,consumer_spent_messages);
    }
  }

  /* If employed then update savings with wage */
  if(agent->firmid != 0)
  {
    agent->savings=agent->savings + agent->wage;
  }
  return 0;
}


/** \fn Spread_awareness()
 * \brief Send message saying location*/
__FLAME_GPU_FUNC__ int Spread_awareness(xmachine_memory_Mall* agent)
{
  //dont need this
  return 0;
```

```
}

/** \fn Job_market()
 * \brief Job market */
__FLAME_GPU_FUNC__ int Job_market(xmachine_memory_Mall* agent,
                 xmachine_message_vacancy_list* vacancy_messages)
{
  int total_vacancies = 0;
  agent->firm_1=0;
  agent->firm_2=0;
  agent->firm_3=0;
  agent->firm_4=0;
  agent->firm_5=0;
  agent->total_vacancies=total_vacancies;

  /* Read vacancy messages */
  xmachine_message_vacancy* current_message =
                 get_first_vacancy_message(vacancy_messages);
  while(current_message)
  {
    /* If the vacancy message refers to this mall*/
    if(agent->id == current_message->mall_id)
    {
      if(agent->firm_1==0)
      {
        agent->firm_1=current_message->firm_id;
        agent->firm_1_vacancy=current_message->vacancies;
        total_vacancies+= current_message->vacancies;
      }
      else if(agent->firm_2==0)
      {
        agent->firm_2=current_message->firm_id;
        agent->firm_2_vacancy=current_message->vacancies;
        total_vacancies+= current_message->vacancies;
      }
      else if(agent->firm_3==0)
      {
        agent->firm_3=current_message->firm_id;
        agent->firm_3_vacancy=current_message->vacancies;
        total_vacancies += current_message->vacancies;
      }
      else if(agent->firm_4==0)
      {
        agent->firm_4=current_message->firm_id;
        agent->firm_4_vacancy=current_message->vacancies;
        total_vacancies += current_message->vacancies;
      }
      else if(agent->firm_5==0)
      {
        agent->firm_5=current_message->firm_id;
        agent->firm_5_vacancy=current_message->vacancies;
        total_vacancies += current_message->vacancies;
      }
      else
      {}
    }
    /* Move onto next message to check */
```

```
    current_message = get_next_vacancy_message(current_message,
                                          vacancy_messages);
  }
  agent->total_vacancies=total_vacancies;
  return 0;
}


__FLAME_GPU_FUNC__ int Job_market_b(xmachine_memory_Mall* agent,
            xmachine_message_application_list* application_messages,
            xmachine_message_employed_list* employed_messages)
{
  int i;
  int total_free_workers = 0;
  int workers[WORKERSNO];
  float wages[WORKERSNO];
  for(i=0;i<WORKERSNO;i++)
  {
    workers[i]=0;
    wages[i]=0.0f;
  }
  int count=0;
  /* Read application messages */
  xmachine_message_application* current_message =
            get_first_application_message(application_messages);
  while(current_message)
  {
    /* If the vacancy message refers to this mall*/
    if(agent->id == current_message->mall_id)
    {
      workers[count]=current_message->person_id;
      wages[count]= current_message->person_wage;
      count++;
      total_free_workers++;
    }
    /* Move onto next message to check */
    current_message = get_next_application_message(current_message,
                                          application_messages);
  }
  /* sort workers cheapest first */
  bubble_sort(&workers, &wages, WORKERSNO);
  count=0;
  while((agent->total_vacancies > 0) && (total_free_workers > 0))
  {
  /* Loop though firm list */
  if(total_free_workers > 0)
  {
    /* If firm still has vacancies */
    if(agent->firm_1_vacancy > 0)
    {
      add_employed_message(employed_messages,workers[count],
                wages[count], agent->firm_1, MESSAGE_RANGE, agent->x,
                                          agent->y, 0.0f);
      //remove_int(&workers, 0);
      //remove_float(&wages, 0);
      total_free_workers--;
      agent->total_vacancies--;
      agent->firm_1_vacancy= agent->firm_1_vacancy - 1;
```

```
        count++;
      }
      else if(agent->firm_2_vacancy > 0)
      {
        add_employed_message(employed_messages,workers[count],
                  wages[count], agent->firm_2, MESSAGE_RANGE, agent->x,
                                                    agent->y, 0.0f);
        total_free_workers--;
        agent->total_vacancies--;
        agent->firm_2_vacancy= agent->firm_2_vacancy - 1;
        count++;
      }
      else if(agent->firm_3_vacancy > 0)
      {
        add_employed_message(employed_messages,workers[count],
              wages[count], agent->firm_3, MESSAGE_RANGE, agent->x,
                                                    agent->y, 0.0f);
        //remove_int(&workers, 0);
        //remove_float(&wages, 0);
        total_free_workers--;
        agent->total_vacancies--;
        agent->firm_3_vacancy= agent->firm_3_vacancy - 1;
        count++;
        }
        else if(agent->firm_4_vacancy > 0)
        {
          add_employed_message(employed_messages,workers[count],
                wages[count], agent->firm_4, MESSAGE_RANGE, agent->x,
                                              agent->y, 0.0f);
          total_free_workers--;
          agent->total_vacancies--;
          agent->firm_4_vacancy= agent->firm_4_vacancy - 1;
          count++;
        }
        else if(agent->firm_5_vacancy > 0)
        {
          add_employed_message(employed_messages,workers[count],
            wages[count], agent->firm_5, MESSAGE_RANGE, agent->x,
                                              agent->y, 0.0f);
          total_free_workers--;
          agent->total_vacancies--;
          agent->firm_5_vacancy= agent->firm_5_vacancy - 1;
          count++;
        }
        else
        {}
      }
    }
    return 0;
}

/** \fn Goods_market()
 * \brief Goods market*/
__FLAME_GPU_FUNC__ int Goods_market(xmachine_memory_Mall* agent,
            xmachine_message_firm_stock_price_list* firm_stock_price_messages)
{
  int firms[FIRMSNO];
```

```
int stock[FIRMSNO];
float price[FIRMSNO];
int consumers[PEOPLENO];
float spending[PEOPLENO];
int firmcount=0;

/* Read goods stock price messages */
xmachine_message_firm_stock_price* current_message =
           get_first_firm_stock_price_message(firm_stock_price_messages);
while(current_message)
{
  /* If the vacancy message refers to this mall*/
  if(agent->id == current_message->mall_id)
  {
    firms[firmcount]=current_message->firm_id;
    stock[firmcount]=current_message->stock;
    price[firmcount]=current_message->price;
    firmcount=firmcount+1;
  }

  /* Move onto next message to check */
  current_message = get_next_firm_stock_price_message
                      (current_message,firm_stock_price_messages);
}

/* Sort goods cheapest first */
sort_prices(&firms, &stock, &price, firmcount);
agent->firm_1_goods=firms[0];
agent->firm_1_stock=stock[0];
agent->firm_1_price=price[0];
agent->firm_2_goods=firms[1];
agent->firm_2_stock=stock[1];
agent->firm_2_price=price[1];
agent->firm_3_goods=firms[2];
agent->firm_3_stock=stock[2];
agent->firm_3_price=price[2];
agent->firm_4_goods=firms[3];
agent->firm_4_stock=stock[3];
agent->firm_4_price=price[3];
agent->firm_5_goods=firms[4];
agent->firm_5_stock=stock[4];
agent->firm_5_price=price[4];
return 0;
}

__FLAME_GPU_FUNC__ int Goods_market_b(xmachine_memory_Mall* agent,
        xmachine_message_consumer_spending_list* consumer_spending_messages,
        xmachine_message_consumer_spent_list* consumer_spent_messages)
{
  int consumers[PEOPLENO];
  float spending[PEOPLENO];
  int i;
  int can_buy = 1;
  int consumer_count;
  int consumercnt=0;

  /* Read consumer spending messages */
```

```
xmachine_message_consumer_spending* current_message =
          get_first_consumer_spending_message(consumer_spending_messages);
while(current_message)
{
  /* If the vacancy message refers to this mall*/
  if(agent->id == current_message->mall_id)
  {
    if(consumercnt<PEOPLENO)
    {
      consumers[consumercnt]=current_message->person_id;
      spending[consumercnt]=current_message->spending;
      consumercnt++;
    }
  }

  /* Move onto next message to check */
  current_message = get_next_consumer_spending_message(current_message,
                                consumer_spending_messages);
}

/* Sell goods until no consumer can buy more goods */
int firmsnumber=5;
while(can_buy)
{
  /* Count consumers not buying */
  consumer_count = 0;
  /* Loop though consumer list */
  for(i=0; i<consumercnt; i++)
  {
    /* If firms left */
    if(firmsnumber > 0)
    {
      /* If enough spending to buy */
      if(spending[i] >= agent->firm_1_price)
      {
        /* Buy one good */
        agent->firm_1_stock = agent->firm_1_stock - 1;
        spending[i] = spending[i] - agent->firm_1_price;
        /* If zero stock for current firm then dismiss */
      }
      else consumer_count++;
    }
  }
  /* If consumers not buying */
  if(consumer_count == consumercnt) can_buy = 0;
}
/* Loop though consumer list */
for(i=0; i<consumercnt; i++)
{
  add_consumer_spent_message(consumer_spent_messages,consumers[i],
  spending[i], MESSAGE_RANGE, agent->x, agent->y, 0.0f);
}
return 0;
}


__FLAME_GPU_FUNC__ int Goods_market_c(xmachine_memory_Mall* agent,
                   xmachine_message_firm_stock_list* firm_stock_messages)
```

```
{
  /* Loop though any remaining firms */
  if(agent->firm_1_stock == 0)
  {
    add_firm_stock_message(firm_stock_messages,agent->firm_1_goods,
          agent->firm_1_stock, MESSAGE_RANGE, agent->x, agent->y, 0.0f);
  }
  if(agent->firm_2_stock > 0)
  {
    add_firm_stock_message(firm_stock_messages,agent->firm_2_goods,
          agent->firm_2_stock, MESSAGE_RANGE, agent->x, agent->y, 0.0f);
  }
  if(agent->firm_3_stock > 0)
  {
    add_firm_stock_message(firm_stock_messages,agent->firm_3_goods,
          agent->firm_3_stock, MESSAGE_RANGE, agent->x, agent->y, 0.0f);
  }
  if(agent->firm_4_stock > 0)
  {
    add_firm_stock_message(firm_stock_messages,agent->firm_4_goods,
          agent->firm_4_stock, MESSAGE_RANGE, agent->x, agent->y, 0.0f);
  }
  if(agent->firm_5_stock > 0)
  {
    add_firm_stock_message(firm_stock_messages,agent->firm_5_goods,
            agent->firm_5_stock, MESSAGE_RANGE, agent->x, agent->y, 0.0f);
  }
  return 0;
}
#endif // #ifndef _FUNCTIONS_H_

<!-- the 0 starting xml file is the same as FLAME HPC-->
<states>
<itno>0</itno>
<xagent>
  <name>Firm</name>
  <id>20</id>
  <value>10</value>
  <a>10</a>
  <productivity>10</productivity>
  <profits>1</profits>
  <f>0.1</f>
  <production>50</production>
  <goodsproduced>0</goodsproduced>
  <stock>0</stock>
  <sold>0</sold>
  <labour>0</labour>
  <numberofworkers>0</numberofworkers>
  <price>1</price>
  <oldprice>1</oldprice>
  <priceinflation>0</priceinflation>
  <sprice>0</sprice>
  <lprice>0</lprice>
  <workerid>{}</workerid>
  <workerwage>{}</workerwage>
  <avewage>1</avewage>
  <mall_id>{}</mall_id>
```

```
    <mall_vacancy>0</mall_vacancy>
    <mall_goods>0</mall_goods>
    <posx>21.8712</posx>
    <posy>36.9246</posy>
</xagent>
<xagent>
    <name>Person</name>
    <id>21</id>
    <savings>10</savings>
    <wage>10</wage>
    <firmid>0</firmid>
    <mall_application>0</mall_application>
    <mall_shopping>0</mall_shopping>
    <mall_id>{}</mall_id>
    <posx>36.9133</posx>
    <posy>-40.3836</posy>
</xagent>
<xagent>
    <name>Mall</name>
    <id>123</id>
    <app_person_ids>{}</app_person_ids>
    <app_person_wages>{}</app_person_wages>
    <sell_firm_ids>{}</sell_firm_ids>
    <sell_firm_stocks>{}</sell_firm_stocks>
    <posx>-14.6247</posx>
    <posy>23.8174</posy>
</xagent>
</states>
```

9.1.1 Visualizing Is Easy in FLAME GPU

Working with the graphics processing unit and CUDA libraries, FLAME GPU comes with built-in tags, which can be used to draw Graphical User Interfaces (GUIs) to help control the simulation and interact it with in real-time. Following is an abstract from a pedestrian model which allows sliders to be attached with global variables. These sliders can allow values to be changed in real-time, and immediately affects the simulation running on the screens.

```
<!-- Within the Environment tag-->
<gpu:environment>
<gpu:constants>
    <gpu:variable>
        <type>float</type><name>COLLISION_WEIGHT</name>
        <control_label>Wall Collision force</control_label>
        <control>Slider</control>
        <group_name>Forces</group_name>
        <limit_high>1.0</limit_high>
        <limit_low>0.0</limit_low>
        <increment>0.1</increment>
        <default>0.5</default>
```

```
    </gpu:variable>
    <gpu:variable>
      <type>float</type><name>GOAL_WEIGHT</name>
      <control_label>Goal force move towards exit</control_label>
      <control>Slider</control>
      <group_name>Forces</group_name>
      <limit_high>1.0</limit_high>
      <limit_low>0.0</limit_low>
      <increment>0.05</increment>
      <default>0.03</default>
    </gpu:variable>
    <gpu:variable>
      <type>int</type><name>SHOPPING_TIME_LIMIT</name>
      <control_label>Time person spends in shop</control_label>
      <control>Text</control>
      <group_name>Times</group_name>
      <limit_high></limit_high>
      <limit_low></limit_low>
      <increment>5</increment>
      <default>30</default>
    </gpu:variable>
    <gpu:variable>
      <type>float</type><name>MOVING_SPEED</name>
      <control_label>Speed of Family</control_label>
      <control>Slider</control>
      <group_name>Speeds</group_name>
      <limit_high>10.0</limit_high>
      <limit_low>0.0</limit_low>
      <increment>0.5</increment>
      <default>0.15</default>
    </gpu:variable>

    <gpu:functionFiles>
      <file>functions.c</file>
    </gpu:functionFiles>
  </gpu:environment>
```

Along with the global variable, agents can be defined with variables which will be visible for plotting on the GUIs.

```
<gpu:xagent>
  <name>pedestrian</name>
  <memory>
  <gpu:variable><type>int</type><name>id</name></gpu:variable>
  <gpu:variable mtransfer="true"><type>float</type><name>x</name></gpu:variable>
  <gpu:variable mtransfer="true"><type>float</type><name>y</name></gpu:variable>
  <gpu:variable><type>float</type><name>velx</name></gpu:variable>
  <gpu:variable><type>float</type><name>vely</name></gpu:variable>
```

```
<gpu:variable><type>float</type><name>speed</name></gpu:variable>
<gpu:variable><type>int</type><name>group_id</name></gpu:variable>
<gpu:variable mtransfer="true"><type>int</type><name>is_leader</name>
  </gpu:variable>
<gpu:variable><type>int</type><name>age</name></gpu:variable>
<gpu:variable mtransfer="true"><type>int</type><name>agent_type</name>
  </gpu:variable>
<gpu:variable><type>int</type><name>exit_no</name></gpu:variable>
</memory>

<functions>
<gpu:function>
  <name>output_pedestrian_location</name>
  <currentState>p_one</currentState>
  <nextState>p_one</nextState>
  <outputs>
    <gpu:output>
    <messageName>pedestrian_location</messageName>
    <gpu:type>single_message</gpu:type>
    </gpu:output>
  </outputs>
  <gpu:reallocate>false</gpu:reallocate>
  <gpu:RNG>false</gpu:RNG>
</gpu:function>

<gpu:function>
  <name>group_forces</name>
  <currentState>p_one</currentState>
  <nextState>p_two</nextState>
  <inputs>
    <gpu:input><messageName>pedestrian_location</messageName></gpu:input>
  </inputs>
  <condition>
    <lhs><agentVariable>function_chosen</agentVariable></lhs>
    <operator>==</operator>
    <rhs><value>0</value></rhs>
  </condition>
  <gpu:reallocate>false</gpu:reallocate>
  <gpu:RNG>true</gpu:RNG>
</gpu:function>
<!--....more functions-->
</functions>

<states>
<gpu:state>
  <name>p_one</name>
  <gpu:visualize>
    <gpu:model><gpu:path>people/man.mesh.xml</gpu:path></gpu:model>
    <gpu:material><name>AgentMat</name></gpu:material>
    <gpu:skeleton>
      <gpu:path>people/man.skeleton.xml</gpu:path>
      <gpu:weightsPerVertex>1</gpu:weightsPerVertex>
      <gpu:maxBlend>1</gpu:maxBlend>
      <gpu:maxAnim>8</gpu:maxAnim>
      <gpu:maxFrame>12</gpu:maxFrame>
    </gpu:skeleton>
    <gpu:transferVars>
```

```
      <gpu:transferVar>x</gpu:transferVar>
      <gpu:transferVar>y</gpu:transferVar>
      <gpu:transferVar>steer_x</gpu:transferVar>
      <gpu:transferVar>steer_y</gpu:transferVar>
      <gpu:transferVar>agent_type</gpu:transferVar>
    </gpu:transferVars>
  </gpu:visualize>
</gpu:state>
<initialState>p_one</initialState>
</states>
<gpu:type>continuous</gpu:type>
<gpu:bufferSize>65536</gpu:bufferSize>
</gpu:xagent>
```

9.1.2   Utilizing Vector Calculations

Through CUDA libraries, complex mathematical functions can be easily defined in FLAME GPU. Those, particularly adapted for graphical representations in certain kinds of models, can benefit from this. For example, pedestrian or biological models which use scene analysis can benefit by being written in FLAME GPU. Economic models, however, can be analyzed as graphs. In these situations HPC simulations can allow more agents to be modeled and simulated. Thus both FLAME versions can be suited for particular models being investigated.

```
collision_force*=COLLISION_WEIGHT;
goal_force*=GOAL_WEIGHT;
total_force.x=collision_force.x+goal_force.x;
total_force.y=collision_force.y+goal_force.y;
agent->x+=total_force.x;
agent->y+=total_force.y;
```

9.2   Commercial Applications of FLAME

FLAME is currently being used for a number of innovative commercial applications. These are exploiting the predictive capabilities of agent-based modeling in two main areas - the management of hospital services and in the management of people in built environments such as passengers in transport hubs (stations, airports etc) and shoppers in retail malls and more.

Patient flow decision support system (by Hao Bai, Mike Holcombe, Laura Smith) Managing emergency departments (ED) has been a challenge for a long time. Since the 4-hour time target was introduced by the UK Government, many EDs have struggled to meet it. It's widely accepted that process management is one of the key reasons. Using the agent-based modeling techniques, a decision support system was developed to obtain information ahead of time in order to optimize the management of human and material resources. By modeling each patient, members of staff and other related resources as agents, this model has the nature of flexibility and capability of providing any detail. With the benefit of modularized design, this platform can also be easily customized for any ED.

Decision support system: The system model is based upon the detailed set of agent types which have been developed based on not only observations at the hospital but also from the use of historical data going back three years. The decision support system is composed of a web-based interface and the ABM model running at the back-end. The architecture of the proposed system is displayed in Figure 9.1. The system could connect with the real-time patient record database for presenting the current status of the whole department. Upon user's request, when a prediction is needed, a snapshot could be generated, and from there a faster than real-time predictive simulation would be run based on historical data. The predictive results would be provided to users to enable them to look ahead of time. If an intervention is necessary, users would be able to apply the intervention to the model, and re-run the simulation to test its effectiveness.

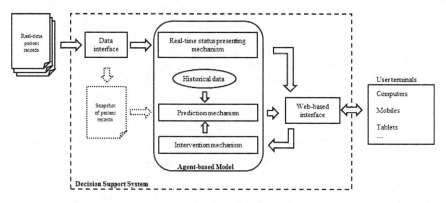

FIGURE 9.1: Diagram of ABM-based decision support system.

For example, on arrival in the green zone reception (self-admitting patients) the receptionist agent will collect information from the patient and when this happens the patient agent will be given an *id* and a basic set of details which are then 'attached' to it in the model as attributes. The patient then waits in the waiting area unless they are identified as an emergency. The next stage will be triage and for this a nurse, when available, will examine the patient and record the basic aspects of triage.

This information is again attached to the patient but will also be broadcast for other staff to read when appropriate, e.g. when making decisions about who a doctor should see next or during further treatment. The patient flow in the green zone is presented in Figure 9.2. The patient agents go through the flow and interact with staff agents at each process. Depending on the availability of staff and other resources, the performance of the green zone is therefore simulated and the model naturally generates outcome based on randomness.

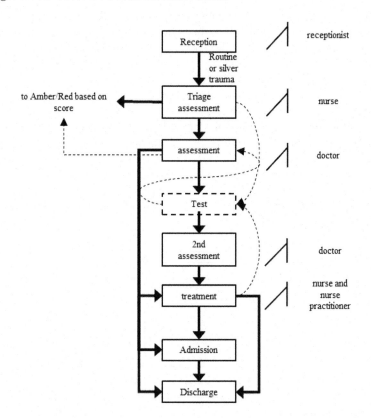

FIGURE 9.2: Patient flow in green zone versus resource usage.

As a state-of-the-art approach, the agent-based model captures the complexities of the process of patient flow in the ED and provides an opportunity to monitor the situation to predict what is likely to happen in the next 4, 12, 24, 48 or more hours. This gives managers and clinicians a foresight view to understand where the blockages are and optimize their management accordingly. Future work will be carried out to collect more information and data from CMFT ED for the purpose of further model testing and validation, which will improve the realism and accuracy of this system. Real-time patient records and staff measurement data will be connected to the model to enable the model to present on-going status of the department and to predict from any chosen scenario.

Concoursia active management of crowds (by Mark Burkitt, Twin Karmakharm, Mike Holcombe).

The management of transport hubs and other places where people gather is becoming increasingly complex. Managing these places is a challenge as safety and security in the light of sudden changes in circumstances are critical. Ensuring that they operate in the most efficient way is paramount.

Overcrowding in stations is a common problem that hits the headlines with commuters seething at mainline stations because of being evacuated, rush hours or overcrowding. It is not just in stations; airports can have problems due to flight delays, weather, terrorist attacks and other emergencies. This gives rise to a number of questions:

> "How can we decide what is the best thing to do? How can we inform passengers about what is going on if we don't know ourselves? What about other places where people gather?"

Problems may arise due to,

- Building works alterations: Closing channels, exits, parts of terminals etc.

- Weather-related problems: Snow, flooding

- Seasonal surges in traffic: Bank holidays, Christmas, strikes

- Managers have to make decisions to prevent overcrowding: How can we help this?

Concoursia is based on the FLAME GPU version (Figure 9.3). It utilizes the ability to represent pedestrians and passengers as agents and imports graphics data to describe the environment where people are interacting. The agents can avoid obstacles and each other, travel in groups, have routes and paths and behave independently with random bizarreness, e.g. as drunken passengers.

The system provides very rapid simulations, faster than real-time and can be used for active management, planning new developments and alterations to existing facilities. For example, coupled with suitable sensor technology, the system can identify where each individual is and the direction they are travelling in to provide predictions of the building state in the next 1030 minutes. If there are signs of potential overcrowding situations, then different available intervention plans can be explored using a rapid simulation capability (Figure 9.4).

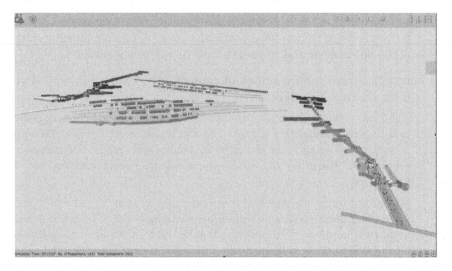

FIGURE 9.3: Part of a Concoursia simulation of a London main station.

It might be useful to have heat maps and other outputs to understand the situation better.

In terms of planning activities, the system can be used to understand the impact of retail activities on passenger behavior, such as how passengers interact with shops? Not every visitor is travelling, so the impact of passenger behavior on retail potential is of interest. Is the retail potential of the transport hub maximized by the location and mix of retail activities (are the shops in the right place)? The impact of flows generated by retail activities can be studied? Are they compatible or conflicting with flows related to the main function of the airport as a transport hub with associated stations, bus stands, car parks etc.? It is common for temporary exhibitions/events to be placed on the concourse, so what is their impact on pedestrian flow? What are the best dates/times for them to be held? (Figure 9.5)

There has been much recent research on how people behave in environments. It can predict what people will do in timescales from 10 minutes to several hours. This is based on either direct feeds from sensors or historical data. It can be used to test out scenarios or interventions so that managers can understand what the best strategy is. In Concoursia, each individual is simulated within the building and predict their movements. Information can be presented in many ways and managers can use this for

- Planning new developments
- Dealing with changes to the layout of the airport

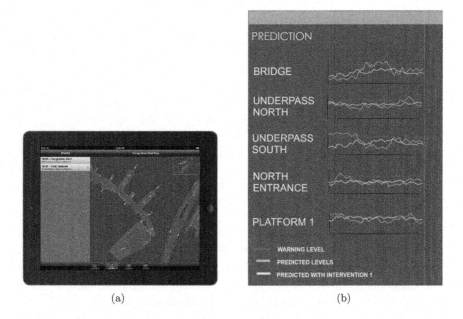

(a)   (b)

FIGURE 9.4: A heat map of the station showing overcrowding (red) and a graphical output over time.

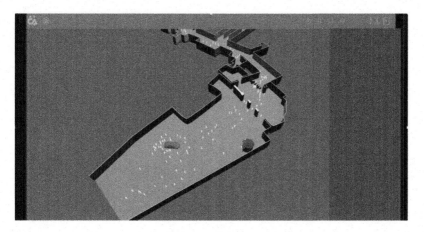

FIGURE 9.5: Modeling the impact of a car display and a pop-up kiosk in a station concourse.

- Planning for emergencies
- Understanding retail opportunities
- Telling passengers what is happening

In planning new developments, extensions can be modeled to see how they might work before expensive building works. How will these changes affect the continuing use of the building? Understanding these changes such as closing/opening channels, routes, exits, facilities and more is important.

Active management applications of Concoursia will involve the integration of suitable sensor technology to feed into the system, such as Wifi, CCTV and more. These are all useful sources of information about where people are and where they are going. Dealing with overcrowding, emergencies and incidents is achieved through rapid simulation of the effects of possible choices to choose best interventions. If there is an incident, then reviewing the issues and carrying out audits after the event is possible with data saved by Concoursia. Thus companies can analyze and improve performance, responses and maintain detailed records of activity for review, key performance indicators and more.

The Concoursia system allows for better management of places where many people gather. It can provide the basis for active safe management of these spaces and greatly improve customer satisfaction.

Bibliography

[1] S. Adra, M. Kiran, P. McMinn, and N. Walkinshaw. A multiobjective optimisation approach for the dynamic inference and refinement of agent-based model specifications. *IEEE Congress on Evolutionary Computation (CEC)*, 2011.

[2] S. Adra, S. Tao, S. MacNeil, M. Holcombe, and R. Smallwood. Development of a three dimensional multiscale computational model of the human epidermis. *PLoS One*, 5(1), 2010.

[3] A. Al, A. E. Eiben, and D. Vermeulen. An experimental comparison of tax systems in sugarscape. *CIEF*, 2000. Online: http://www.cs.vu.nl/~gusz/papers/CIEF2000-Al-Eiben-Vermeulen.ps.

[4] R. Albert, H. Jeong, and A. L. Barabasi. Error and attack tolerance of complex networks. *Nature*, 406(378), 2000.

[5] F. Alkemade. *Evolutionary Agent-Based Economics*. PhD thesis, Institute of Programming Research and Algorithms, 2004. Part of project 'Evolutionary Systems for Electronic Markets'.

[6] C. Altavilla, L. Luini, and P. Sbriglia. Information and learning in Bertrand and Cournot experimental duopolies. Economics Working Paper Series 406, University of Siena, October 2004.

[7] M. Altaweel, N. Collier, T. Howe, R. Najlis, M. North, M. Parker, E. Tatara, J. R. Vos, L. Girardin, and L. Gulyas. Repast: Recursive Porus Agent Simulation Toolkit, 2005. Online: http://repast.sourceforge.net/index.html.

[8] P. W. Anderson. More is different: Broken symmetry and the nature of the hierarchical structure of science. *Science*, 177:393–396, 1972.

[9] J. Arifovic. Genetic algorithm learning and the cobweb model. *Journal of Economic Dynamics and Control*, (18):3–28, 1994.

[10] J. Arifovic. The behavior of exchange rate in the genetic algorithm and experimental economics. *Journal of Political Economy*, 104(3):510–541, 1996.

[11] W. B. Arthur, S. N. Durlauf, and D. Lane. *The Economy as an Evolving Complex System II.* Addison Wesley, 1997.

[12] R. Axelrod. The evolution of strategies in the iterated prisoner's dilemma. *Genetic Algorithms and Simulated Annealing*, pages 32–41, 1987.

[13] R. Axelrod. *The complexity of Cooperation: Agent-Based Models of Competition and Collaboration.* Princeton University Press, Princeton, NJ, 1997.

[14] Robert Axelrod. *The Evolution of Cooperation.* Basic Books, New York, 1984.

[15] T. Bäck, F. Hoffmeister, and P. Schwefel. A survey of evolution strategies. *Proceedings of the Fourth International Conference on Genetic Algorithms*, pages 2–9, 1991.

[16] T. Bäck, D. Vermeulen, and A. E. Eiben. Tax and Evolution in Sugarscape. Online: http://www.cs.vu.nl/~gusz/papers/Tax-and-evolution.ps.

[17] H. Bai, M. D. Rolfe, W. Jia, S. Coakley, R. K. Poole, J. Green, and M. Holcombe. Agent-based modeling of oxygen-responsive transcription factors in escherichia coli. *PLoS Computational Biology*, 10(4), 2014.

[18] Mark J. Baldwin. A new factor in evolution. *The American Naturalist*, 30(354):441–451, June 1896.

[19] J. Barr and F. Saraceno. Cournot competition, organisation and learning. *Journal of Economics Dynamics and Control*, 29:277–295, 2005.

[20] K. Beck, M. Beedle, A. Bennekum, A. Cockburn, W. Cunningham, M. Fowler, J. Grenning, J. Highsmith, A. Hunt, R. Jeffries, J. Kern, B. Marick, R. C. Martin, S. Mellor, K. Schwaber, J. Sutherland, and D. Thomas. Principles behind the agile manifesto. *Agile Alliance*, 2001.

[21] E. Beinhocker. *The Origin of Wealth: Evolution, Complexity and the Radical Remaking of Economics.* Random House Business Books, April 2007.

[22] J. Bertrand. Book review of theorie mathematique de la richesse sociale and of recherches sur les principes mathematiques de la theorie des richesses. *Journal de Savants*, 67:499–508, 1883.

[23] M. Bicak. *Agent-Based Modeling of Decentralized Ant Behavior Using High Performance Computing.* PhD thesis, University of Sheffield, Sheffield, UK, 2011.

[24] Susan J. Blackmore. *The Meme Machine.* Oxford University Press, Oxford, Oxfordshire, UK, 1999.

[25] E. Bonabeau, M. Dorigo, and G. Theraulaz. *Swarm Intelligence: From Natural to Artificial Systems.* Oxford University Press, Oxford, 1999.

[26] A. H. Bond and L. Gasser. A survey of distributed artificial intelligence. In Alan H. Bond and Les Gasser, editors, *Readings in Distributed Artificial Intelligence.* Morgan Kaufmann Publishers, 1988. San Mateo, CA, Online: http://www.exso.com/nsurvo.pdf.

[27] G. Bontempi and Y. Le Borgne. An adaptive modular approach to the mining of sensor network data. In *Proceedings of the Workshop on Data Mining in Sensor Networks, SIAM SDM,* pages 3–9. SIAM Press, 2005.

[28] A. Bosch-Domenech and N. J. Vriend. Imitation of successful behavior in cournot markets. Economics Working Papers 269, Department of Economics and Business, Universitat Pompeu Fabra, February 1998. Online: http://ideas.repec.org/p/upf/upfgen/269.html.

[29] Ester Böserup. *The Conditions of Agricultural Growth: The Economics of Agrarian Change under Population Pressure.*

[30] H. J. Bremermann. Optimisation through evolution and recombination. *Self-Organising Systems,* pages 93–106, 1962.

[31] R. A. Brooks. Intelligence without reason. *Proceedings of the Twelfth International Joint Conference on Artificial Intelligence (IJCAI-91),* pages 569–595, 1991.

[32] R. A. Brooks. Intelligence without representation. *Artificial Intelligence,* 47:139–159, 1991.

[33] Mark Buchanan. Why economic theory is out of whack. New Scientist, July 2008. ¡Online:http://www.newscientist.com/article/mg19926651.700?DCMP=ILC-rhts&nsref=ts1_bar¿.

[34] M. Burkitt. *Computational Modeling of Sperm Behavior in a 3D Virtual Oviduct.* PhD thesis, University of Sheffield, Sheffield, UK, 2011.

[35] M. Burkitt, M. Kiran, S. Konur, M. Gheorghe, and F. Ipate. Agent-based high-performance simulation of biological systems on the gpu. *IEEE International Conference on High Performance Computing and Communications,* 2015.

[36] P. Buzing, A. Eiben, and M Schut. Emerging communication and cooperation in evolving agent societies. *Journal of Artificial Societies and Social Simulation,* 8(1-2), 2005. Online: http://jasss.soc.surrey.ac.uk/8/1/2.html.

[37] S. Camazine, J. L. Deneubourg, N. R. Franks, J. Sneyd, G. Theraulaz, and E. Bonabeau. *Self-Organization in Biological Systems*. Princeton University Press, Princeton, NJ, 2001.

[38] Lewis Caroll. *Through the Looking-Glass: And What Alice Found There*. A. L. Burt, New York, 1897. Drawings by John Tenniel.

[39] S. Chin, D. Worth, C. Greenough, S. Coakley, M. Holcombe, and M. Kiran. The EURACE agent-based economic model - benchmarking, assessment and optimization. *RAL Technical Reports RAL-TR-2012-006*, 2012.

[40] S. Coakley. *Formal Software Architecture for Agent-Based Modeling in Biology*. PhD thesis, Department of Computer Science, University of Sheffield, Sheffield, UK, 2007.

[41] S. Coakley, M. Gheorghe, M. Holcombe, L. S. Chin, D. Worth, and C. Greenough. Exploitation of high performance computing in the FLAME agent-based simulation framework. *Proceedings of the 14th International Conference on High Performance Computing and Communications*, 2012.

[42] S. Coakley, R. Smallwood, and M. Holcombe. From molecules to insect communities - how formal agent-based computational modeling is uncovering new biological facts. *Scientifae Mathematicae Japonicae Online*, pages 765–778, March 2006.

[43] E. G. Coffman, M. Elphick, and A. Shoshani. System deadlocks. *Computing Surveys*, 2:67–78, 1971.

[44] A. Cournot. *Researches into the Mathematical Principles of the Theory of Wealth*. Macmillan, New York, 1897.

[45] R. Cressman. *The Stability Concept of Evolutionary Game Theory*. Number 94 in Lecture Notes in Biomathematics. Springer-Verlag, Berlin, 1992.

[46] R. K. Dash, N. R. Jennings, and D. C. Parkes. Computational-mechanism design: A call to arms. *IEEE Computer Society*, 18(6):40–47, November/December 2003.

[47] H. Dawid. *Adaptive l=Learning by Genetic Algorithms: Analytical Results and Applications to Economic Models*. Springer-Verlag, Berlin, 1999.

[48] R. Dawkins. *The Selfish Gene, 2nd edition*. Oxford University Press, Oxford, 1989.

[49] R. H. Day. *Adaptive processes and economic theory, Adaptive Economic Models*, pages 1–38. Academic Press Inc., New York, 1975.

[50] R. Dellinger. Roguelike intelligence - genetic algorithms and evolving state machine AIs. Roguelike Development, July 2009. Online: `http://roguebasin.roguelikedevelopment.org/index.php?title=Roguelike_Intelligence_-_Genetic_Algorithms_and_Evolving_State_Machine_AIs`.

[51] M. d'Inverno and M. Luck. *Understanding Agent Systems.* Springer-Verlag, Berlin, 2001.

[52] K. Dopfer. *Economics in the Future,* pages 3–35. Macmillan, Tokyo, Japan, 1976.

[53] M. Dorigo. *Optimisation, Learning and Natural Algorithms.* PhD thesis, Politecnico di Milano, Italy, 1992.

[54] D. C. Dracopoulos. *Evolutionary Learning Algorithms for Neural Adaptive Control, Perspectives in neural computing.* Springer, 1997.

[55] J. Duffy. Learning to speculate: Experiments with artificial and real agents. *Journal of Economic Dynamics and Control,* 25:295–319, 2001.

[56] J. Epstein. *Generative Social Science: Studies in Agent-Based Computational Modeling.* Princeton University Press, Princeton, NJ, January 2007.

[57] J. M. Epstein and R. Axtell. *Growing Artificial Societies: Social Science from the Bottom Up.* MIT Press, Cambridge, MA, 1996.

[58] M. D. Ernst, J. H. Perkins, P. J. Guo, S. McCamant, C. Pacheco, M. S. Tschantz, and C. Xiao. The daikon system for dynamic detection of likely invariants. *Science of Computer Programming,* 69(1-3):35–45, December 2007.

[59] P. A. Fiskwick. Simulation Model Design and Execution: Building Digital Worlds. *Prentice-Hall,* 1995.

[60] D. B. Fogel. Evolving behaviors in the iterated prisoner's dilemma. *Evolutionary Computation,* 1(1):77–97, 1993.

[61] D. B. Fogel. *Evolutionary Computation: Toward a New Philosophy of Machine Intelligence.* IEEE Press, USA, 1995.

[62] L. J. Fogel. Autonomous automata. *Industrial Research,* 4:14–19, 1962.

[63] L. J. Fogel. Decision making by Automata. Technical Report GDA-ERR-AN-222, General Dynamics, San Diego, CA, 1962.

[64] L. J. Fogel. *On the Organisation of Intellect.* Phd dissertation, University of California, Los Angeles, CA, 1964.

[65] L. J. Fogel. On the design of conscious automata. Final Report AF 49(638)-1651, AFOSR, Arlington, VA, 1966.

[66] L. J. Fogel and G. H. Burgin. Competitive goal-seeking through evolutionary programming. Final Report AF19(628)-5927, Air Force Cambridge Research Labs, 1969.

[67] S. A. Frank. *Foundations of Social Evolution.* Princeton University Press, Princeton, New Jersey, 1998.

[68] R. M. Friedberg. A learning machine: Part 1. *IBM Journal of Research and Development*, 2:2–13, 1958.

[69] G. J. Friedman. Digital simulation of an evolutionary process. *General Systems Yearbook*, 4:171–184, 1959.

[70] M. Friedman. *Essays in Positive Economics.* University of Chicago Press, Chicago, 1953.

[71] G. Fullstone, J. Wood, M. Holcombe, and G.iuseppe Battaglia. Modeling the transport of nanoparticles under blood flow using an agent-based approach. *Scientific Reports*, 5(10649), 2015.

[72] M. Gardner. Mathematical games. *Scientific American*, October 1970.

[73] N. Gilbert and P. Terna. How to build and use agent-based models in social science. *Mind and Society*, 1(1):1860–1839, May 1999. Online: http://www.springerlink.com/content/8878861366x5q522/fulltext.pdf.

[74] J. Goldstein. Emergence as a construct: History and issues. *Emergence: Complexity and Organisation*, 1:49–72, 1999.

[75] M. Granovetter. *Economic action and social structure: the problem of embeddedness, The Sociology of Economic Life*, pages 51–76. Westview Press, USA, 2001.

[76] C. Greenough, L. S. Chin, D. Worth, S. Coakley, M. Holcombe, and M. Kiran. An approach to the parallelization of agent-based applications. *ERCIM News*, 81:42–43, 2010.

[77] V. Grimm, U. Berger, F. Bastiansen, S. Eliassen, V. Ginot, J. Giske, J. Goss-Custard, T. Grand, S. K. Heinz, G. Huse, A. Huth, J. U. Jepsen, C. Jorgensen, W. M. Mooij, B. Müller, G. Péer, C. Piou, S. F. Railsback, A. M. Robbins, M. M. Robbins, E. Rossmanith, N. Rüger, E. Strand, S. Souissi, R. A. Stillman, R. Vabø, U. Visser, and D. L. DeAngelis. A standard protocol for describing individual-based and agent-based models. *Ecological Modeling, Elsevier*, 198:115–126, June 2006. Online: http://www.ufz.de/data/GrimmODD4780.pdf.

[78] W. Gropp, E. Lusk, D. Ashton, P. Balaji, D. Buntinas, R. Butler, A. Chan, J. Krishna, G. Mercier, R. Ross, R. Thakur, and B. Toonen. Mpich2 user's guide version 1.0.5. *Argonne National Laboratory, Chicago*, 2006.

[79] G. J. Gumerman, A. C. Swedlund, J. S. Dean, and J. M. Epstein. The evolution of social behavior in the prehistoric american southwest. *Artificial Life*, 9:435–444, 2003. Online: http://www.mitpressjournals.org/doi/pdf/10.1162/106454603322694861.

[80] D. Hales. Memetic engineering and cultural evolution. In L. Doglas Kiel, editor, *Knowledge Management, Organisational Intelligence and Learning, and Complexity*, Encyclopedia of Life Support Systems (EOLSS), Oxford, UK, 2004. Developed under the auspices of the UNESCO, Eolss Publishers. http://www.eolss.net.

[81] P. G. Harrald and D. B. Fogel. Evolving continuous behaviors in the iterated prisoner's dilemma. *Biosystems*, 37:135–145, 1996.

[82] R. L. Haupt and S. E. Haupt. *Practical Genetic Algorithms*. Wiley, 2004.

[83] X. He, A. Prasad, S. P. Sethi, and G. Guthierrez. A survey of stackelberg differential game models in supply and marketing channels. *Journal of Systems Science and Systems Engineering (JSSSE)*, 16(4):385–413, December 2007. Online: http://papers.ssrn.com/sol3/papers.cfm?abstract_id=1069162.

[84] A. E. Henniger. Emotional synthetic forces. Technical Report 1149, United States Army Research Institute for Behavioral and Social Sciences, 2004. Online: http://www.au.af.mil/au/awc/awcgate/army/tr1149.doc.

[85] J. Hofbauer and K. Sigmund. *Evolutionary Games and Population Dynamics*. Cambridge University Press, Cambridge, 1998.

[86] M. Holcombe, S. Adra, M. Bicak, L. S. Chin, S. Coakley, A. Graham, J. Green, C. Greenough, D. Jackson, M. Kiran, S. MacNeil, A. Maleki-Dizaji, P. McMinn, M. Pogson, R. Poole, E. Qwarnstrom, F. Ratnieks, M. Rolfe, R. Smallwood, S. Tao, and D. Worth. Modeling complex biological systems using an agent-based approach. *Integrative Biology*, 4(1):53–46, 2012.

[87] M. Holcombe, S. Coakley, M. Kiran, S. Chin, C. Greenough, D. Worth, S. Cincotti, M. Raberto, A. Teglio, C. Deissenberg, S. van der Hoog, H. Dawid, S. Gemkow, P. Harting, and M. Neugart. Large-scale modeling of economic systems. *Complex Systems*, 22(2):175–191, 2013.

[88] J. H. Holland. Adaptive plans optimal for payoff-only environments. *Proceedings of the Second Hawaii International Conference on System Sciences*, pages 917–920, 1969.

[89] J. H. Holland. *Adaptation in Natural and Artificial Systems*. University of Michigan Press, Ann Arbor, 1975.

[90] J. H. Holland and J. H. Miller. Artificial adaptive agents in economic theory. *AEA Papers and Proceedings*, May 1991.

[91] J. J. Hopfield. Physics, biology and complementarity. In D. de Boer, E. Dahl, and O. Ulfbeck, editors, *The Lesson from Quantum Theory, Proceedings of the Niels Bohr Centenary Symposium*, 1996.

[92] D. Horres and R. Gore. Exploring the similarities between economic theory and artificial societies. National Science Grant 0426971, University of Virginia, department of Computer Science, Virginia, 2008. Online: http://www.cs.virginia.edu/~rjg7v/pubs/capstone_08_horres_rgore.pdf.

[93] L. Hurwicz, E. S. Maskin, and R. B. Myerson. Mechanism design theory. *Sveriges Riksbank Prize in Economic Sciences in Memory of Alfred Nobel 2007*, 2007. Online: http://nobelprize.org/nobel_prizes/economics/laureates/2007/ecoadv07.pdf.

[94] A. Ilachinski. *Cellular Automata*. World Scientific Publishing, Singapore, 2001.

[95] P. A. Ioannou and A. Pitsillides. *Modeling and Control of Complex Systems*. CRC Press, Los Angeles, CA, 2007.

[96] D. Jackson, M. Bicak, and M. Holcombe. Decentralized communication, trail connectivity and emergent benefits of ant pheromone trail networks. *Memetic computing*, 3(1):25–32, 2011.

[97] D. Jackson, M. Holcombe, and F. Ratnieks. Trail geometry gives polarity to ant foraging networks. *Nature*, 432(7019):907–909, 2004.

[98] S. Johnson. *Emergence: The Connected Lives of Ants, Brains, Cities and Software*. Penguin Press, UK, 2001.

[99] D. Kahneman. *Thinking, Fast and Slow*. 2011.

[100] M. I. Kamien and N. L. Schwartz. *Dynamic Optimisation: The Calculus of Variations and Optimal Control in Economics and Management*. Elsevier, North Holland, 1991.

[101] E. R. Kandel and L. R. Squire. Neuroscience: Breaking down scientific barriers to the study of brain and mind. *Science*, 290(1113), 2000.

[102] A. Kannankeril. Sugarscape-growing agent-based artificial societies, July 1996. Online: `http://sugarscape.sourceforge.net/`.

[103] S. A. Kaufmann. *The Origins of Order*. Oxford University Press, New York, 1993.

[104] P. Kefalas, G. Eleftherakis, M. Holcombe, and M. Gheorghe. Simulation and verification of P systems through communicating X-machines. *BioSystems*, 70(2):135–148, 2003.

[105] J. G. Kemeny. Man viewed as machine. *Scientific American*, 192:58–67, April 1955.

[106] M. Kim and A. Petersen. An evaluation of Daikon: A dynamic invariant detector. *Daikon Report*, 2004.

[107] M. Kiran, S. Coakley, N. Walkinshaw, P. McMinn, and M. Holcombe. Validation and discovery from computational biology models. *Biosystems*, 93(1-2):141–150, July-August 2008.

[108] M. Kiran, P. Richmond, M. Holcombe, L. S. Chin, D. Worth, and C. Greenough. FLAME: Simulating large populations of agents on parallel hardware architectures. *AAMAS, Toronto, Canada*, May 2010.

[109] D. Klöck. Extended sugarscape model in XL. Technical report, Department of Graphic Systems, University of Cottbus. Online: `http://danielkloeck.wdfiles.com/local--files/extended-sugarscape/StudienarbeitSugarscape.pdf`.

[110] J. Koza, M. A. Keane, M. J. Streeter, W. Mydlowec, J. Yu, and G. Lanza. *Genetic Programming IV Routine Human-Competitive Machine Intelligence*. Kluwer Academic Publishers, 2003.

[111] J. E. Laird, C. Congdon, and K. J. Coulter. *SOAR User's Manual: Version 8.6.3*. Electrical Engineering and Computer Science Department, University of Michigan, October 2006.

[112] J. E. Laird, A. Newell, and P. S. Rosenbloom. SOAR: An architecture for general intelligence. *Artificial Intelligence*, 33(1):1–64, 1987.

[113] J. B. Lamarck. Philosophie zoologique, ou exposition des considrations relatives à l'historic naturelle des animaux, Paris, 1809.

[114] C. G. Langton. Computation at the edge of chaos. *Physica D*, 42, 1990.

[115] B. LeBaron. Empirical regularities from interacting long and short horizon investors in an agent-based stock market model. *IEEE transaction on Evolutionary computation*, 5:442–455, 2001.

[116] R. Lewis. *Complexity - Life at the Edge of Chaos*. J. M. Dent, London, 1993.

[117] X. Li, A. K. Upadhyay, A. J. Bullock, T. Dicolandrea, J. Xu, R. L. Binder, M. K. Robinson, D. R. Finlay, K. J. Mills, C. C. Bascom, C. K. Kelling, R. J. Isfort, J. W. Haycock, S. MacNeil, and R. Smallwood. Skin stem cell hypotheses and long term clone survival explored using agent-based modeling. *Scientific Report*, 3(1904), 2013.

[118] Ambient Life. Sugarscape. ALife, 2009. Online: `http://www.ambientlife.net/alife/asoc.html`.

[119] N. Lin. Social Capital: A Theory of Social Structure and Action. *Cambridge University Press*, 2001.

[120] W. Liu and M. Niranjan. The role of regulated mRNA stability in establishing bicoid morphogen gradient in drosophila embryonic development. *PLoS ONE*, 6(9), 2011.

[121] W. Liu and M. Niranjan. Gaussian process modeling for bicoid mRNA regulation in spatiotemporal bicoid profile. *Bioinformatics*, 28(3):366–372, 2012.

[122] E. N. Lorenz. Chaos: In view of the inevitable inaccuracy and incompleteness of weather observations, precise very-long-range forecasting would seem to be non-existent. *Journal of the Atmospheric Sciences*, 20:130–141, 1962. Online: `http://complex.upf.es/~josep/Chaos.html`.

[123] S. Luke, L. Panait, G. Balan, S. Paus, Z. Skolicki, E. Popovici, J. Harrison, J. Bassett, R. Hubley, and A. Chircop. Ecj 16: A Java-based Evolutionary Computation Research System, 2007. Online: `http://cs.gmu.edu/~eclab/projects/ecj/`.

[124] N. Magnenat and D. Thalmann. *Groups and Crowd Simulation*. Wiley, 2004.

[125] A. Maleki-Dizaji, M. Holcombe, M. D. Rolfe, P. Fisher, J. Green, R. K. Poole, A. I. Graham, and SysMO-SUMO Consortium. A systematic approach to understanding escherichia coli responses to oxygen: From microarray raw data to pathways and published abstracts. *Online Journal of Bioinformatics*, 10:51–59, 2009.

[126] M. Malthus. *An Essay on the Principle of Population*. Norton Critical Editions, 1803.

[127] R. E. Marks. Breeding hybrid strategies: Optimal behavior for oligopolists. *Journal of Evolutionary Economics*, 2:17–38, 1992.

[128] P. Maunder, D. Myers, N. Wall, and R. Miller. *Economics explained: A coursebook in A Level Economics*. Collins Educational, London, UK, 1989.

[129] E. Mayr. *Toward a New Philosophy of Biology: Observations of an Evolutionist.* Belknap Press, Cambridge, MA, 1988.

[130] D. McFadden. Consumer choice behavior and the measurement of well-being. *The New Science of Pleasure*, December 2012.

[131] J. H. Miller and S. E. Page. *Complex Adaptive Systems: An Introduction to Computational Models of Social Life.* Princeton University Press, May 2007.

[132] John H. Miller. The coevolution of automata in the repeated prisoner's dilemma. Technical Report 89-003, SFI Economics Research Program, Sante Fe Institute and Carnegie-Mellon University, 1989.

[133] M. Minsky. Steps toward artificial intelligence. *Proceedings of the IRE*, 49(1):8–30, 1961.

[134] M. Minsky. *The Society of Mind.* Simon and Schuster, New York, 1985.

[135] M. Minsky. *Understanding Musical Activities: Readings in A.I. and the Music (Laske Interview).* AAAI Press, Menlo Park, CA, May 1991. Online: http://web.media.mit.edu/~minsky/papers/Laske.Interview.Music.txt.

[136] P. Monge and N. Contractor. Communication networks: Measurement techniques. *Handbook for the Study of Human Communication*, 33:107–138, 1988.

[137] A. Narguney. *Sustainable Transportation Networks.* Edward Elgar Publishing, Cheltenham, England, 2000.

[138] J. Nash. Equilibrium points in n-person games. *Proceedings of the National Academy of Sciences*, 36(1):48–49, 1950.

[139] J. Nash. Non-cooperative games. *The Annals of Mathematics*, 54(2):286–295, 1951.

[140] A. Newell. *Unified Theories of Cognition.* Harvard Press, Cambridge, MA, 1990.

[141] V. Nieminen. Game of Life Simulation. Online: http://vesanieminen.com/img/portfolio/gameoflife.png, 2009.

[142] J. Noble. Cooperation, conflict and the evolution of communication. *Journal of Adaptive Behavior*, 7(3-4):349–370, 1999.

[143] E. Norling. Contrasting a system dynamics model and an agent-based model of food web evolution. *Lecture Notes in AI*, (4442):57–68, 2007.

[144] E. Ott, T. Sauer, and J. A. Yorke. *Coping with Chaos: Analysis of Chaotic Data and the Exploitation of Chaotic System.* Wiley-Interscience, August 1994.

[145] Ovid. *Metamorphoses: Book the First.* Sir Samuel Garth and John Dryden. Online: http://classics.mit.edu/Ovid/metam.1.first.html.

[146] D. Parkes. *Iterative Combinatorial Auctions: Achieving Economic and Computational Efficiency.* PhD thesis, Department of Computer and Information Science, University of Pennsylvania, May 2001.

[147] L. J. Peter and R. Hull. *The Peter Principle: Why Things Always Go Wrong.* William Morrow and Company, New York, 1969. OCLC, WorldCat.

[148] I. Peterson. The gods of sugarscape: Digital sex, migration, trade and war on the social science frontier. *Science News*, 150(21):332–336, November 1996. Online: http://www.sciencenews.org/pages/pdfs/data/1996/150-21/15021-18.pdf.

[149] M. Pogson, M. Holcombe, R. Smallwood, and E. Qwarnstrom. Introducing spatial information into predictive nfκb modeling - an agent-based approach. *PLOS ONE*, 3(6), June 2008. Online: doi:10.1371/journal.pone.0002367.

[150] W. Poundstone. Prisoner's Dilemma Doubleday. 1992.

[151] C. Prell and M. Kiran. The processes of social capital and the emergence of network structure. *Proceedings from Sunbelt XXX. Riva del Garda Fierecongressi, Trento, Italy*, 2010.

[152] T. C. Price. Using coevolutionary programming to simulate strategic behavior in markets. *Journal of Evolutionary Economics*, (7):219–254, 1997.

[153] S. F. Railsback, S. L. Lytinen, and S. K. Jackson. Agent-based simulation platforms: Review and development recommendations. *Simulation*, 82(9):609–623, 2008. Online: http://sim.sagepub.com/cgi/content/abstract/82/9/609.

[154] A. Rapoport. Optimal policies for the prisoner's dilemma. NIH Grant 50, Psychimetric Laboratory, University of North Carolina, 1966. MH-10006.

[155] I. Rechenberg. Cybernetic solution path of an experimental problem. Library Translation 1122, Royal Aircraft Establishment, August 1965.

[156] I. Rechenberg. *Evolutionsstrargie: Optimerung Technisher Systeme nach Prinzipien der Biologischen Evolution.* Fromman Holzboog Verlag, Stuttgart, 1973.

[157] F. Vega Redondo. *Evolution, Games and Economic Behavior.* Oxford University Press, Oxford, 2008.

[158] J. Reed, R. Toombs, and N. A. Barricelli. Simulation of biological evolution and machine learning. *Journal of Theoretical Biology,* 17:319–342, 1967.

[159] P. Rendall. A Turing Machine in Coney's Game of Life. 2001. Online: www.cs.ualberta.ca/~builtko/F02/papers/tm_words.pdf.

[160] C. Reynolds. Flocks, herds, and schools: A distributed behavior model. *ACM Siggraph,* 1987.

[161] D. M. Rhodes, S. A. Smith, M. Holcombe, and E. E. Qwarnstrom. Computational modeling of nfkb activation by il-1ri and its co-receptor tilrr, predicts a role for cytoskeletal sequestration of ikba in inflammatory signalling. *PLoS ONE,* 10(6), 2015.

[162] P. Richmond, S. Coakley, and D. Romano. A high performance agent-based modeling framework on graphics card hardware with CUDA. *Proceedings of 8th International Conference on Autonomous Agents and Multiagent Systems,* 2009.

[163] E. Robinson, M. Holcombe, and F. Ratnieks. The organization of soil disposal by ants. *Animal Behavior,* 75(1389), 2008.

[164] E. Robinson, D. Jackson, M. Holcombe, and F. Ratnieks. Insect communication: 'No entry' signal in ant foraging. *Nature,* 438(7067):442, 2005.

[165] A. Rogers, R. K. Dash, S. D. Ramchurn, P. Vytelingum, and N. R. Jennings. Coordinating team players within a noisy iterated prisoners dilemma tournament. *Theoretical Computer Science,* 377(1-3):243–259, 2007.

[166] S. H. Roosta. *Parallel Processing and Parallel Algorithms, Theory and Computation.* Springer, New York, 2000.

[167] P. Ross and D. Corne. Comparing genetic algorithms, simulated annealing and stochastic hill climbing on timetabling problems. In *Evolutionary Computing,* volume 993 of *Lecture Notes in Computer Science,* pages 94–102. Springer-Berlin/Heidelberg, January 1995.

[168] C. T. Rubin, J. Seater, J. Dike, and C. Weed. Letters. *Science News,* 151(3):45–46, January 1997. Online: http://www.sciencenews.org/pages/pdfs/data/1997/151-03/15103_02.pdf.

[169] Thomas C. Schelling. Dynamic models of segregation. *Journal of Mathematical Sociology,* 1:143–186, 1971.

[170] R. Schoonderwoerd, O. Holland, J. Bruten, and L. Rothkrantz. Ant-based load balancing in telecommunication networks. *Adaptive Behavior*, 5(2):169–207, 1997.

[171] H. Schuster. *Complex Adaptive Systems: An Introduction.* Scator Verlag, Germany, 2001.

[172] H. P. Schwefel. *Kybernetische Evolution als Strategie der Experimentellen Forschung in der Strömungstechnik.* PhD thesis.

[173] H. P. Schwefel. *Numerical Optimisation of Computer Models.* John Wiley, Chichester, 1981.

[174] Giovanna Di Marzo Serugendo, Anthony Karageorgos, Omer F. Rana, and Franco Zambonelli, editors. *Engineering Self-Organising Systems Nature-Inspired Approaches to Software Engineering*, number 2977 in Lecture Notes in Artificial Intelligence. Springer, 2004.

[175] R. E. Shannon. Systems Simulation: The Art and Science. *Prentice-Hall*, 1975.

[176] Y. Shoham. Agent-oriented programming. Technical Report 133590, Department of Computer Science, Stanford University, Stanford, CA, 1990.

[177] K. Sigmund. *Games of Life.* Oxford University Press, Oxford, 1993.

[178] G. Silverberg. Evolutionary modeling in economics: Recent history and immediate prospects. *Evolutionary Economics as a Scientific Research Programme*, May 1997.

[179] Herbert A. Simon. *The Sciences of Artificial Worlds.* MIT Press, 1998.

[180] K. Sims. Evolving 3D morphology and behavior by competition. *Artificial Life IV Proceedings*, 1994. Online: http://www.lri.fr/~devert/homepage/bibliography.html.

[181] K. Sims. Evolving virtual creatures. *SIGGRAPH 1994 Proceedings*, pages 15–22, 1994.

[182] A. Sloman. The SimAgent TOOLKIT – for philosophers and engineers (also known as SIM_AGENT), May 2005.

[183] A. Sloman. Your Mind: CogAff Project, 2007. Online: http://www.cs.bham.ac.uk/ axs/fig/your.mind.jpg.

[184] A. Sloman and B. Logan. Architectures and tools for human-like agents. In *Proceedings of the Second European Conference on Cognitive Modeling, ECCM-98*, pages 58–65. Nottingham University Press, 1998.

[185] A. Smith. *The Theory of Moral Sentiments.* Edinburgh, 1759.

[186] J. Maynard Smith. *Evolution and the Theory of Games.* Cambridge University Press, December 1982.

[187] J. Maynard Smith. *Evolutionary Genetics.* Oxford University Press, 1989.

[188] J. Maynard Smith and G. R. Price. The logic of animal conflict. *Nature,* (246):15–18, November 1973.

[189] J. Maynard Smith and E. Szathmary. *The Major Transitions in Evolution.* W.H. Freeman/Spektrum, Hiedelberg, Oxford, New York, 1995.

[190] T. Snijders, C. Steglich, and G. van de Bunt. Introduction to actor-based models for network dynamics. *Social Networks,* 33:44–60, 2010.

[191] Lionhead Studios, Electronic Arts, Feral Interactive, and Microsoft Coorporation. Black and White, 2001. Online: http://www.lionhead.com/bw/.

[192] S. Tao, P. McMinn, M. Holcombe, R. Smallwood, and S. MacNeil. Agent-based modeling helps in understanding the rules by which fibroblasts support keratinocyte colony formation. *PLoS ONE,* 3(5), 2008.

[193] L. Tesfatsion. Agent-Based Computational Economics, January 2007. Online: http://www.econ.iastate.edu/tesfatsi/ace.htm.

[194] L. Tesfatsion and R. Axelrod. *Handbook of Computational Economics: Agent-Based Computational Economics,* volume 2 of *Handbooks in Economic series.* Elsevier/North Holland, Amsterdam, 2006. Online: http://www.econ.iastate.edu/tesfatsi/abmread.htm.

[195] Alan M. Turing. Computing machinery and intelligence. *Mind,* (59):433–460, 1950.

[196] L. Valen. A new evolutionary law. *Evolutionary Theory,* 1:1–30, 1973.

[197] S. Varma. *Hybrid Hierarchical Computational Models of Cardiac Cells and Tissues.* PhD thesis, University of Sheffield, Sheffield, UK, 2010.

[198] J. von Neumann. *The Theory of Self-Reproducing Automata.* University of Illinois Press: A. Burks, Illinois, 1966.

[199] J. von Neumann and O. Morgenstern. *Theory of Games and Economic Behavior.* Princeton University Press, 1944.

[200] N. J. Vriend. An illustration of the essential difference between individual and social learning, and its consequences for computational analyses. *Journal of Economic Dynamics and Control,* 24:1–19, 2000.

[201] The free encyclopedia Wikipedia. Agent-Based Model, 2007. Online: http://en.wikipedia.org/wiki/Agent-based_model.

[202] E. O. Wilson and B. Hölldobler. Dense heterarchies and mass communications as the basis of organisation in ant colonies. *Trends in Ecology and Evolution*, 3:65–68, 1988.

[203] S. Wolfram. *A New Kind of Science*. Wolfram Media (May 14, 2002), 2002.

[204] M. Wooldridge. *The Logical Modeling of Computational Multi-Agent Systems*. PhD thesis, Department of Computer Science, UMIST, Manchester, UK, 1992.

[205] M. Wooldridge and N. R. Jennings. Intelligent agents: Theory and practice. *The Knowledge Engineering Review 10*, 2:115–152, 1995.

[206] J. B. Xavier and K. R. Foster. Cooperation and conflict in microbial biofilms. *PNAS*, 104(3):876–881, January 2007.

[207] Z. Zhang and C. Zhang, editors. *Agent-Based Hybrid Intelligent Systems, an Agent-Based Framework for Complex Problem Solving*, number 2938 in Lecture Notes in Artificial Intelligence. Springer, 2004.

Index

agent models, 14
agent-based frameworks, 33
agile methods, 48
ant modeling, 202

bounded rationality, 125

cancer modeling, 224
Chaos theory, 4
Complex adaptive systems, 3
Concoursia, 279
Cournot model, 129

dynamic system, 6

emergence, 6
equilibrium, 174
event-driven, 28
evolutionary computation, 18
extreme programming, 51

FLAME, 34, 43
FLAME GPU, 247

JADE, 37

Libmboard, 54

Markov modeling, 40
MASON, 37
Modeling and simulation, 10

Netlogo, 36

pedestrian modeling, 114
Prisoner's dilemma game, 164

reinforcement learning, 26
Repast, 36

Schelling's model, 88
SimAgent, 35
SOAR, 35
social networks, 107
Sugarscape model, 89
SWARM, 33
synchronization, 57

time-driven, 28
Turing machines, 19

unit testing, 237

X-machines, 44